# Contemporary chemistry for schools and colleges

Written by Vanessa Kind
RSC School Teacher Fellow 2001-2002

KT-433-798

ROYAL SOCIETY OF CHEMISTRY

# Contemporary chemistry for schools and colleges

Written by Vanessa Kind

Edited by Colin Osborne and Maria Pack

Designed by Russel Spinks and Alberto Arias at russel-spinks.co.uk

Published and distributed by Royal Society of Chemistry

Printed by Royal Society of Chemistry

Copyright © Royal Society of Chemistry 2004

Registered charity No. 207890

Apart from any fair dealing for the purposes of research or private study, or criticism or review, as permitted under the UK Copyright Designs and Patents Act, 1988, this publication may not be reproduced, stored or transmitted, in any form or by any means, without the prior permission in writing of the publishers, or in the case of reprographic reproduction, only in accordance with the terms of licences issued by the appropriate Reproduction Rights Organisation outside the UK. Enquiries concerning reproduction outside the terms stated here should be sent to the Royal Society of Chemistry at the London address printed on this page.

While every effort has been made to contact owners of copyright material, we apologise to any copyright holders whom we have unwittingly infringed.

For further information on other educational activities undertaken by the Royal Society of Chemistry contact:
Email: *education@rsc.org*
Telephone: 020 7440 3344

Education Department
Royal Society of Chemistry
Burlington House
Piccadilly
London
W1J 0BA

Information on other Royal Society of Chemistry activities can be found on its websites:
*www.rsc.org*
*www.chemsoc.org*
*www.chemsoc.org/LearnNet* contains resources for teachers and students from around the world.

ISBN 0 - 85404 - 382 - 9

British Library Cataloguing in Publication Data.

A catalogue for this book is available from the British Library.

# Foreword

Chemistry is the central science in the modern world. It subsumes all of molecular physics and molecular biology and the new areas of nanoscience and nanotechnolgy are actually subdisciplines of chemistry. Without a detailed understanding of chemistry at a fundamental level, today's biologists, physicists, material scientists, nanoscientists and many others would be able to do little worthwhile.

This resource contains material from many areas where chemistry is important and is described at a level which can be understood by 11–16 year olds. Teachers should be encouraged to use these contexts wherever possible in their curriculum-based presentations as they will help to enthuse and motivate their students.

Prof Sir Harry Kroto CChem Hon FRSC FRSE FRS

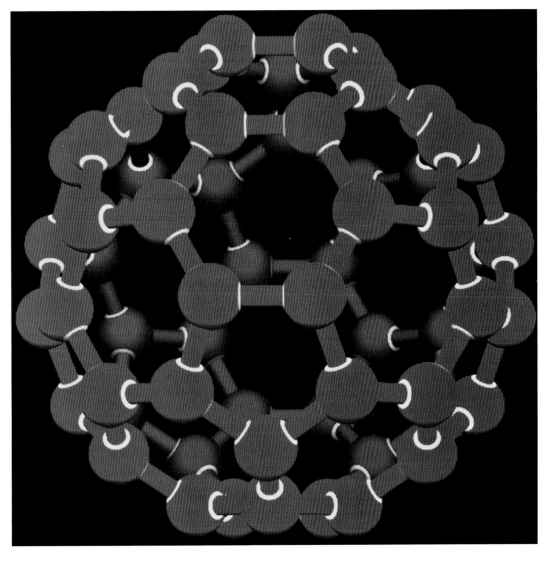

# Acknowledgements

This project could not have been written without the help of many people around the world who contributed time and resources to the preparation of these materials.

The author would also like to thank the following people for their help in producing this resource.

Colin Osborne and Maria Pack of the RSC Education department for their support and tolerance during the writing process.
Penny Andrews, The Energy Saving Trust, UK
Stephen Atkinson, PA Photos Limited, UK
Chris Baker, Education Consultant, Cheshire, UK
Rahila Begum, Neals Yard Remedies, UK
Charles Blue, National Radio Astronomy Observatory, AUI, Virginia, USA
Peter Borrows, CLEAPSS, Uxbridge, UK
David Bowers, AWE, Blacknest, Hampshire, UK
Jane Bridge, Canterbury Girls Grammar School, Kent, UK
Lucy Brock, *www.akg-images.co.uk* (accessed November 2003)
Virginia Campbell, General Motors of Canada Limited, Canada
Danise Coon, Chile Pepper Institute, NMSU
Till Credner, Allthesky.com
John Crompton, Procter and Gamble Technical Centres Limited, Surrey, UK
Thomas Dame, Harvard-Smithsonian CfA, Cambridge, Massachusetts, USA
Michael Davidson, The Florida State University, USA
Rosemary Davies, BMW, UK
Cees Dekker, University of Delft, Netherlands
Jane Dietrich, California Institute of Technology, USA
Michael Douma, *www.webexhibits.org/causesofcolour* (accessed November 2003)
Patrick Dunn, Broadstairs Grammar School, Kent, UK
Darryl Flemming, Kimbolton Fireworks, Kimbolton, Cambridgeshire, UK
Ford Motor Company Limited, UK
Romy Fraser, Neals Yard Remedies, London, UK
Arthur Garforth, UMIST, UK
Catherine Gerst, L'Oreal Recherche, France
M. Reza Ghadiri, The Scripps Research Institute, USA
Chris Gummer, Procter and Gamble, UK
Joseph Gurman, NASA Goddard Space Flight Center, Solar Physics Branch, USA
Jonathan Hare, Department of Chemistry, University of Sussex, UK
Dap Hartmann, Leiden Observatory, Netherlands
Pauline Hili, Product Development Manager, Neals Yard Remedies, London UK
Jane Hoffman, Backyard Scientist, USA
R. E. Hubbard, Department of Chemistry, University of York, UK
Milos Kalab, Canada
Frank Katch, Santa Barbara, CA, USA
Terry Kienzle, Gemcolor.com
Angela Kight, Science and Technology Group, Institute of Education, University of London, UK
Per Morten Kind, Department of Physics, NTNU, Trondheim, Norway
Harry Kroto, Department of Chemistry, University of Sussex , UK
Mat Lawrence, MLE Pyrotechnics Limited, Northamptonshire, UK
Kenneth Libbrecht, California Institute of Technology, USA
David Leisawitz, NASA Goddard Space Flight Center
Liz Loeffler, Department of Earth Sciences, University of Bristol
David Malin, Anglo Australian Telescope, Australia
Axel Mellinger, University of Potsdam, Germany
Peter Marshall, AWE, Blacknest, Hampshire, UK
James Maynard, Chemistry Department, University of Wisconsin, USA
Ray Miglionico, Jericho Historical Society, USA

# RS•C Contemporary chemistry for schools and colleges

## Acknowledgements (continued)

Adam Phelps, CfBT, UK
Patricia Phillips-Batoma, Department of Chemistry, University of Illinois at Urbana-Champaign, USA
Maren Pink, Indiana University Molecular Stucture Centre, USA
Valerio Pirronello, Department of Physics, University of Catania, Sicily, Italy
Louise Potter, Johnson Matthey Fuel Cells, Reading, UK
Steve Price, Air Force Research Laboratory, Hanscom, Massachusetts, USA
John Raffan, Department of Education, University of Cambridge, UK
Patricia Rasmussen
Stephanie Richmond, *www.internestantiques.com* (accessed November 2003)
Alain Rochefort, École Polytechnique de Montréal, Canada
Mike Ross, IBM Research Center, Almaden, California, USA
Royal Navy Submarine Museum Photograph Archive, Hampshire, UK
John Rummel, Madison Metropolitan School District, Madison Wisconsin, USA
Aaron Schmidt, Boston Public Library, Boston, USA
Alethea Tabor, Department of Chemistry, University College London, UK
Keith Taber, Department of Education, University of Cambridge, UK
Les Talbot
Paul Tierney, *www.fireworks.co.uk* (accessed November 2003)
The Times Newspapers, London, UK
University of Rochester, New York, USA
Vermont Historical Society, Champlain, Vermont, USA
Serena Viti, Department of Astronomy, University College London, UK
Julia Walton, Department of Chemistry, University of York, York, UK
Jason Ware, *www.galaxyphoto.com* (accessed November 2003)
Dick Waymark, Department of Chemistry, University College London, UK
Samantha Williams, Toyota, UK
J.M. Williams-French, *www.hurstmereclose.freeserve.co.uk* (accessed November 2003)
John Woodford, Michigan Today, USA
Grant Yoxon, Canadian Driver, Canada
David Ford, Castle View School, Canvey Island, UK

### Images

Many people have generously contributed images to this resource and are acknowledged in the appropriate place in the text/student sheet. We do not hold the copyright for many of these images and therefore it is strictly forbidden to download images without permission from the original source.

The Royal Society of Chemistry would like to extend its gratitude to Professor Donald McIntyre and the Department of Education, University of Cambridge and Peter van Marion and the Program for Lærerutdanning at the Norwegian University of Science and Technology, Trondheim, for providing office accommodation for this Fellowship and Professors Michael Reiss and Geoff Whitty, Institute of Education, University of London, for seconding Vanessa Kind to the RSC's Education department.

# Contents

# Contents (continued)

# Contents (continued)

# How to use this resource

This resource has been designed with flexibility in mind. A whole unit could be used, but often a single lesson could be based on a selection of the material, or images or interactive activities could be used to support a lesson or part of a scheme of work.

The teachers' guide contains a full listing of the resource. It includes: the title; the type of activity; the possible age range; the topic and the type of media available.

An annotated page from the teachers' guide is shown on page viii.

All the material except the teacher's guide is available on the CDROM. Photocopiable, printable and projectable materials are available in three formats:
• coloured pdf
• black and white pdf
• Word document (which can be altered or edited by the teacher prior to printing).

Each worksheet has an index number *eg* 01.06 which represents Chapter 1, Worksheet 6. These three formats allow teachers flexibility in using the resource according to their local circumstances. The interactive exercises are on the CDROM and can be saved to a stand-alone computer or a network.

If using black and white copies of the student worksheets, teachers will need to use the projectable picture resources available on the CDROM, where indicated with this symbol:

As the science curriculum changes we hope teachers will use these materials to address the criticism (from Ofsted and others) that contemporary science is not often introduced into the classroom, and that the variety of activities will promote active learning among students.

## CDROM instructions and system requirements

This CDROM is compatible with Windows NT/2000/ME & XP.
Recommended minimum specification: 64MB of RAM.

Insert the CDROM into the CDROM drive. Your PC should run the CDROM automatically. If it does not, select your CDROM drive, using Windows Explorer or My Computer.

You will need a:
Web Browser – the content has been optimised for Internet Explorer 6
Macromedia Flash Player – for the interactive content.
Acrobat PDF Reader – to use the PDF files in the Teacher Zone
Microsoft Word – to use any Word documents in the Teacher Zone.

The CDROM will install or prompt you to install the Flash Player and Acrobat PDF Reader.
It is the responsibility of the user to have MS Word software installed on their system before running this CDROM.

The CDROM licence allows the files on the CDROM to be downloaded and to be accessible over a network. The RSC will not offer support or guidance on how best to network the files.

**Password access to the Teacher Zone on the CDROM**
Enter this *case-sensitive* password when prompted.

RSC2004

## How to use this resource (continued)

### Disclaimer
This CDROM has been thoroughly checked for errors and viruses. The RSC cannot accept liability for any damage to your computer system or data which occurs while using this CDROM or the software contained on it. If you do not agree with these conditions, you should not use the CDROM.

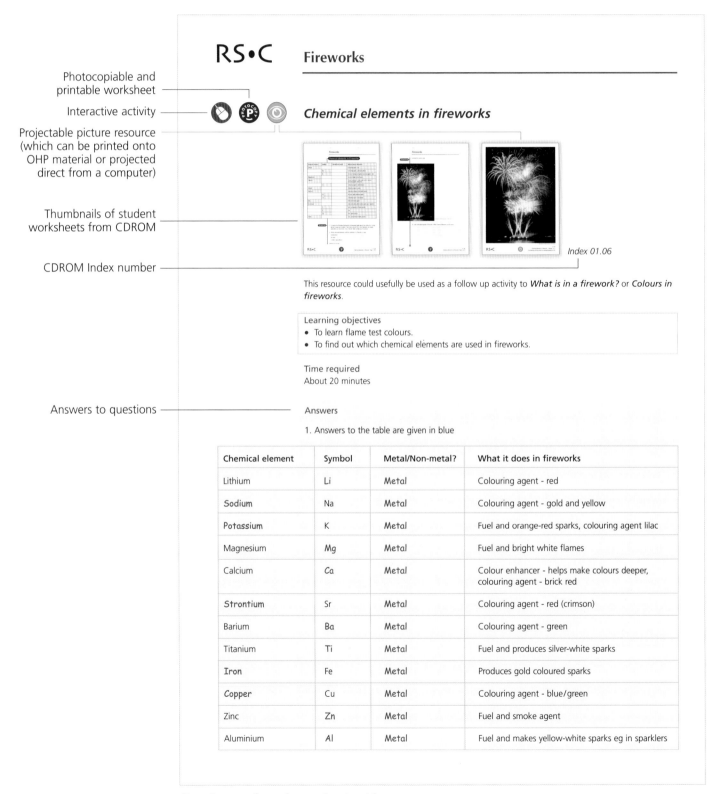

Photocopiable and printable worksheet

Interactive activity

Projectable picture resource (which can be printed onto OHP material or projected direct from a computer)

Thumbnails of student worksheets from CDROM

CDROM Index number

Answers to questions

# RS•C    Fireworks

### *Chemical elements in fireworks*

*Index 01.06*

This resource could usefully be used as a follow up activity to *What is in a firework?* or *Colours in fireworks*.

Learning objectives
- To learn flame test colours.
- To find out which chemical elements are used in fireworks.

Time required
About 20 minutes

Answers

1. Answers to the table are given in blue

| Chemical element | Symbol | Metal/Non-metal? | What it does in fireworks |
|---|---|---|---|
| Lithium | Li | Metal | Colouring agent - red |
| Sodium | Na | Metal | Colouring agent - gold and yellow |
| Potassium | K | Metal | Fuel and orange-red sparks, colouring agent lilac |
| Magnesium | Mg | Metal | Fuel and bright white flames |
| Calcium | Ca | Metal | Colour enhancer - helps make colours deeper, colouring agent - brick red |
| Strontium | Sr | Metal | Colouring agent - red (crimson) |
| Barium | Ba | Metal | Colouring agent - green |
| Titanium | Ti | Metal | Fuel and produces silver-white sparks |
| Iron | Fe | Metal | Produces gold coloured sparks |
| Copper | Cu | Metal | Colouring agent - blue/green |
| Zinc | Zn | Metal | Fuel and smoke agent |
| Aluminium | Al | Metal | Fuel and makes yellow-white sparks eg in sparklers |

*Sample page from the teachers' guide*

# Challenges facing the chemical sciences

Much of the way chemical species (such as molecules, ions, giant structures, and polymers) behave is known and underpinned by theoretical understanding. We can, for example, now manufacture a whole range of materials to have particular properties and make molecules which fit into particular sites in receptors to act as drugs. Good chemistry can make many things possible. However there are still many challenges facing chemists and chemical scientists working in fields such as molecular biology, materials science, nanotechnology etc.

Chemists extend nature to such an extent that 90% of all known substances have been created by chemistry and would not have been known without the efforts of chemists. It is true that in the past not enough care and consideration was given to possible unexpected effects of chemical substances on the environment but new initiatives such as Green Chemistry and the chemical industry's Responsible Care programme are seeking to address this.

Chemistry and the chemical industry contribute hugely to the economic success of the country. The United Kingdom chemical industry contributes over £4 billion to the balance of payments (what we earn abroad compared to what we spend abroad).

Some of the major challenges facing the chemical sciences are:

1. **In medicine**
   Many drugs we use are treatments, not cures. Some diseases as yet have no known cure, we just alleviate the symptoms. The common cold is an everyday, if hardly life threatening, example, whilst ebola is a much more serious case. Chemical scientists need to find cures for these diseases.

2. **In materials**
   We need to be able to produce biocompatible materials in order to be able to replace body parts once and once only. We need superconductors that work at high temperature so the transmission of electricity is much more energy efficient in order to conserve scarce resources.
   We need to be able to create paints that last for decades not years so that structures are painted at the start of their lifetime but never need repainting again (except for aesthetic reasons).

3. **In degradability**
   We need to produce materials that make products that are stable over their life time but are degradable after use.

4. **In molecular biology**
   We need to be able to work out the three dimensional structure of proteins from their amino acid sequence. Whilst we have the sequence in the human genome we cannot yet predict the conformations that proteins can adopt, and without this we cannot treat many diseases.
   We need organised chemical systems that mimic the functions of biological cells to produce molecular machines.

5. **In information and communications technology**
   We need nanoscale molecular computers and transistors to increase computing power even more.

6. **In energy use**
   We need batteries that are small enough and powerful enough to make electric cars a reality. We need ways of dealing with nuclear waste products so that nuclear power can be more widely used.

These challenges need to be addressed by those currently within the school population. They must be enthused to continue with the study of chemistry. The importance of 'blue skies' research (research for its own sake to further extend our knowledge) needs to be emphasised since we cannot predict what novel uses discoveries may be put to. Nobody could have predicted that the discovery

# Challenges facing the chemical sciences (continued)

of light amplification by the stimulated emission of radiation, the laser, would have led to the development of the compact disc, and no industry would have supported such research!

## Teachers' notes

Teachers may find the information above of use when faced with the familiar questions, why we are doing this? What is chemistry for? Why do we need to know about chemistry? and so on.

### 1. GREEN CHEMISTRY

### What is Green Chemistry?

Green Chemistry is the use of chemistry for pollution prevention. More specifically, Green Chemistry is the design of chemical products and processes that reduce or eliminate the use or generation of hazardous substances.

By offering environmentally benign alternatives to the more hazardous chemicals and processes that are often used in both consumer and industrial applications, Green Chemistry is promoting pollution prevention at the molecular level.

### Green Chemistry Focus Areas

Green Chemistry technologies can be categorised into one or more of the following three focus areas:

- the use of alternative synthetic pathways

- the use of alternative reaction conditions

- the design of safer chemicals that are, for example, less toxic than current alternatives or inherently safer with regard to accident potential.

### The 12 Principles of Green Chemistry

1. **Prevention**
   It is better to prevent waste than to treat or clean up waste after it has been created.

2. **Atom economy**
   Synthetic methods should be designed to maximise the incorporation of all materials used in the process into the final product.

3. **Less hazardous chemical syntheses**
   Wherever practicable, synthetic methods should be designed to use and generate substances that possess little or no toxicity to human health and the environment.

4. **Designing safer chemicals**
   Chemical products should be designed to preserve efficacy of function while reducing toxicity.

5. **Safer solvents and auxiliaries**
   The use of auxiliary substances (eg solvents, separation agents, etc.) should be made unnecessary wherever possible and innocuous when used.

# Challenges facing the chemical sciences (continued)

6. **Design for energy efficiency**
   Energy requirements of chemical processes should be recognized for their environmental and economic impacts and should be minimised. If possible, synthetic methods should be conducted at ambient temperature and pressure.

7. **Use of renewable feedstocks**
   A raw material or feedstock should be renewable rather than depleting whenever technically and economically practicable.

8. **Reduce derivatives**
   Unnecessary derivatization (use of blocking groups, protection/deprotection, temporary modification of physical/chemical processes) should be avoided whenever possible.

9. **Catalysis**
   Catalytic reagents (as selective as possible) are superior to stoichiometric reagents.

10. **Design for degradation**
    Chemical products should be designed so that at the end of their function they do not persist in the environment and break down into innocuous degradation products.

11. **Real-time analysis for pollution prevention**
    Analytical methodologies need to be further developed to allow for real-time, in-process monitoring and control prior to the formation of hazardous substances.

12. **Accident prevention**
    Substances and the form of a substance used in a chemical process should be chosen to minimise further the potential for chemical accidents, including releases, explosions, and fires.

## 2.  RESPONSIBLE CARE

Responsible Care is the chemical industry's commitment to continual improvement in all aspects of health, safety and environmental (HS&E) performance and to openness in communication about its activities and its achievements.

The aim of Responsible Care is to earn public trust and confidence through a high level of HS & E performance in order to maintain the industry's licence to continue to operate safely, profitably and with due care for the interests of future generations.

# What is Chemistry and why do we teach it?

Many teachers are asked these questions by their students and feel unsure how to answer. This attempts to answer these questions, and together with the section on the challenges facing chemistry, is designed to help teachers give their students an idea of the place of the subject in the modern world.

Chemistry is the study of substances, what they are made of, how they interact and what role they play in living things. In fact it is the study of all materials and is vital to every aspect of life. From the moment of birth and throughout the whole of life chemistry surrounds us; in the air we breathe, the food we eat and the clothes we wear.

Chemistry is also what chemists do, however this is complicated by the fact that many people not called chemists also do chemistry because it is **the** enabling science. Without it molecular biologists could not sequence the human genome, nanotechnologists could not make molecular computers, and materials scientists could not make superconductors. All of these people depend on chemistry to make new materials and to do this they need to know some chemistry. Chemistry is a subject with a few simple ideas at its core. This resource shows some of these applications.

We teach chemistry for many reasons. Obviously to provide a supply of future chemists, to carry on developing all those things from plastics to composite materials that make our life comfortable. As a by-product of this we also train many people in sought after generic problem solving, analytical, decision-making and numerate skills for a wide variety of careers in banking, commerce and industry. However we also teach chemistry so that those students who will not continue it further will have enough understanding to become effective citizens in an increasingly scientific and technological world. Thus we would hope that such people would be able to see something and realise that what it does depends on the microscopic structure of matter, that much of its behaviour can be explained quantitatively and that chemists have become 'magicians with matter' and can create new substances.

The simple, but abstract, ideas at the core of chemistry are shown overleaf.

# The core of chemistry

1. Matter is composed of tiny particles called atoms and there are about 100 different atoms called elements.

2. Elements show periodic relationships in their chemical and physical properties.

3. These periodic relationships can be explained in terms of their atomic structure (arrangement of neutrons, protons and electrons).

4. Atoms link together (or bond) by either sharing electron pairs or by transferring electrons from one atom to another.

5. The shapes of molecules and the way giant structures are arranged is of great importance in terms of the way they behave. It is the forces between them that determine the bulk properties of matter.

6. Energy is conserved and can therefore be neither created or destroyed, although there is a relationship between energy and mass.

7. Both energy and matter tend to disperse in disorder but this dispersal can generate order.

8. There are barriers to reaction so reactions occur at different rates.

9. Chemical reactions take place in only four different ways:
   - proton transfer
   - electron transfer
   - electron sharing
   - electron pair sharing.

(With grateful acknowledgement to the ideas presented by Professor Peter Atkins, Lincoln College, Oxford in his plenary lecture at the 17th International Conference on Chemical Education, Beijing, 2002 and at the RSC National Meeting for Teachers of Chemistry, Birmingham, 2004.)

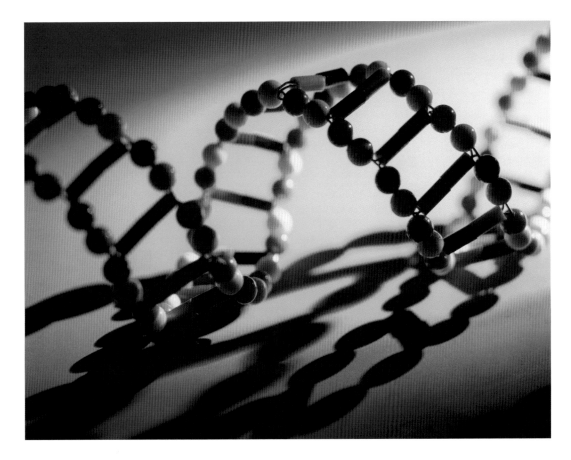

# Fireworks

*Reproduced with kind permission of Paul Tierney.*

# RS•C    Fireworks

**Summary**
Fireworks give us the opportunity to see dramatic, explosive chemistry on a large scale. This unit explores the chemistry behind fireworks and the history of gunpowder.

| Resource name | Index | Type | Age range | Topic | Media |
|---|---|---|---|---|---|
| *What is a firework?* | *01.01* | Experiment | 11–16 | Combustion reactions | |
| *What is in a firework?* | *01.02* | Experiment Interactive activity | 11–16 | Flame test colours and chemical elements in fireworks | |
| *Did you know? About firing fireworks* | *01.03* | Literacy | 11–16 | Firework displays | |
| *Person profile – Darryl Flemming* | *01.04* | Careers information | 11–16 | Careers | |
| *Colours in fireworks* | *01.05* | Demonstration (alternative to **What is in a firework?** class experiment) | 11–16 | Flame test colours and chemical elements in fireworks | |
| *Chemical elements in fireworks* | *01.06* | Paper based exercise Interactive activity Projectable colour picture | 11–16 | Chemical elements in fireworks and the effects they have | |
| *Did you know? About blackpowder* | *01.07* | Literacy | 14–16 | History of blackpowder | |
| *Investigating blackpowder* | *01.08* | Demonstration and questions | 14–16 | Chemistry of gunpowder, explosions and equations | |
| *Making blackpowder* | *01.09* | Literacy comprehension / DART | 14–16 | Manufacture of gunpowder in 1859 | |
| *Key words* | *01.10* | Glossary | 11–16 | Glossary | |

Key:  ⬤ Interactive student activity    Ⓟ Photocopiable and printable worksheet    ◎ Projectable picture resource

Issue

What are fireworks and how do they work?

This material gives the background to fireworks chemistry and sets the making of gunpowder in a historical context.

Chemical topics

- Flame test colours of alkali and other metals
- Characteristics of fireworks
- Chemistry of gunpowder, balancing equations

Scientific enquiry issues

- Scientific work in a historical context

Notes on using the unit

Most students will recognise the traditional English 'poem' about the Gunpowder Plot, which is:

*Remember, remember the 5th of November,*
*Gunpowder, treason and plot*
*There is no good reason why gunpowder treason*
*Should ever be forgot!*

A brief description of the Gunpowder Plot is given at the start of the students' pages and further information can be found by consulting websites listed on page 17. In order to provide a realistic picture of fireworks, a combination of at least two or more of the activities could be used. However, the activities are all free-standing so could be used independently of each other. They are most suited to students aged 11–16. The demonstrations are suitable for use with any age.

In **What is a firework?**, students are encouraged to make careful observations of common events - firing a party popper and a sparkler. These introduce the main effects seen in all fireworks in a safe way. Next, students could go on to find out more about the chemicals producing the colours in fireworks by carrying out flame tests in **What is in a firework?**

An alternative set of instructions is provided for this to be done as a demonstration in **Colours in fireworks**. In **Investigating blackpowder**, demonstration reactions illustrate the nature of explosives, particularly gunpowder. Optional calculations are provided to lead students to the composition of gunpowder. These are appropriate for older students. Note that the information is readily available on the internet, so it is a good idea to be 'up front' with students about responsibility for using chemical information wisely.

**Making blackpowder** looks at making gunpowder in an historical setting, using a fictional account of a child worker in a gunpowder factory to explore the scientific aspects of part of the process. Besides answering the questions, this activity has the potential to make a cross-curricular link to work in history and social studies.

**RS•C** Fireworks

 ## *What is a firework?*

*Index 01.01*

Learning objectives
- To establish the characteristics of fireworks from party poppers and sparklers.
- To learn that setting off a firework is a chemical reaction.

Time required
About 30 minutes

Apparatus and equipment (per group)
- 1 or 2 party poppers
- 1 or 2 sparklers
- Eye protection for each student
- Match/splint to light the sparkler
- Heatproof mat
- Bucket of cold water to dispose of the sparkler
- Copies of the observation table.

Safety
Wear eye protection. Warn students not to fire party poppers at faces and that sparklers will be very hot.

Notes on the experiment
Party poppers are widely available in packs of about 24. Sparklers are available seasonally - in the UK they can be bought in October and at New Year in other European countries.

A firework is an explosive comprising a mixture of substances. The mixture includes fuel, a supply of oxygen and other chemicals to create effects. The party popper and sparkler are simple fireworks which illustrate these principles.

The party popper contains a small amount of explosive. When the string is pulled the explosive is ignited and the contents are fired out. The explosive acts as a lifting charge for the paper streamers inside. The string is like a fuse. The party popper illustrates the basic firework principle of an explosive acting as a lifting charge for other material.

A sparkler illustrates colour and sparks. The sparkler mixture does not contain an explosive. Usually the mixture comprises iron filings, aluminium powder, barium nitrate and a binding agent such as dextrin or gum arabic. Sparklers are made by dipping iron wire repeatedly into the mixture, which is prepared as a wet slurry. Each layer is dried before adding the next. Often huge bundles of sparklers are dipped together. Some sparklers have a primer painted on to the end to make ignition easier. The chemical reaction in a sparkler is:-

$$10Al + 3Ba(NO_3)_2 \rightarrow 3BaO + 3N_2 + 5Al_2O_3$$

4

The barium nitrate acts as a stabiliser and as a source of oxygen to oxidise the aluminium. The reaction is exothermic. The gas produced causes the iron filings to be ejected. The heat from the reaction heats the iron filings so they appear as sparks. The wire conducts the heat and helps to maintain the burn of the mixture.

In the UK all fireworks are graded – party poppers are graded 1, which means that they are suitable for indoor use. 'Garden' fireworks are graded 2 or 3. Display fireworks are usually graded 4 or 5. The grading depends on the power of the explosive mixture. This is defined as the Net Explosive Mass (NEM), a figure which indicates how much of the firework's mass is due to explosive alone.

Observation table

Students may suggest:

| Firework | What I saw | What I smelt and heard |
| --- | --- | --- |
| Party popper | Flash or spark<br>Smoke<br>Paper streamers coming out<br>Base of popper blown away | 'Fireworks' smell / smell of caps like in a gun burning<br>'Bang' or 'pop' sound |
| Sparkler | Sparks in streams<br>Bright yellow light<br>Sparkler melting and dropping down<br>Solid stuff disappearing | 'Metallic' burning smell<br>Fizzing noise |

Answers

1. The effects of the party popper are caused by the explosive lifting the paper streamers out of the container.

2. The solid substance on the sparkler reacted in a chemical reaction. Gases were made in the reaction which went into the air. Iron was in the sparkler. The filings were heated up and also went into the air.

3. The effects of the sparkler are caused by the heat from the chemical reaction. These cause the iron filings to become hot and go into the air as sparks.

4. The main characteristics of fireworks are: colour, light, lifting charge, heat and smoke.

Conclusion - Answer

A suggested answer could be:
A firework is an explosive comprising a mixture of substances where the mixture includes fuel, a supply of oxygen and other chemicals to create effects.

# *What is in a firework?  and Colours in fireworks*

Two alternatives are given here - a class experiment **What is in a firework?** and a demonstration **Colours in fireworks**. Both achieve the same learning objectives. An interactive activity where students 'drag and drop' chemical element names and flame test pictures is available on the CDROM.

Learning objectives
- To learn flame test colours
- To find out which chemical elements are used in fireworks.

Time required
About 50 minutes for the class experiment **What is in a firework?**.
About 30 minutes for the demonstration **Colours in fireworks**.

## *What is in a firework?*

*Index 01.02*

Apparatus and equipment (per group)
- Periodic Table
- Results tables
- Samples of firework chemicals – see list below
- Flame test wires
- Distilled water
- Small beaker / test-tube
- Bunsen burner
- Heatproof mat
- Eye protection.

Chemicals to test
A suggested list may include:-
sodium chloride
potassium chloride
copper(II) carbonate – **Harmful**
strontium chloride
calcium chloride – **Irritant**
barium chloride – **Toxic**
lithium chloride – **Irritant**
zinc chloride – **Corrosive**
magnesium chloride
iron filings.

Chlorides are best to use. The substances can be prepared in petri dishes, labelled on base and top with the name and formula. Preparing two of each would ensure more rapid circulation around the group. Wires should be kept with each dish to avoid contamination. It is advised that teachers demonstrate barium and zinc salts as they are toxic.

 Safety
Wear eye protection.

Notes on the experiment

The class experiment requires students to carry out a flame test on a range of different chemicals. Students should ensure sticks do not become contaminated, keeping sticks with the dishes. The water is used to dampen the end so solid can attach a little more easily. Students often think that far more solid is needed than necessary - encourage frugality! The best colours can be seen by placing the compounds at the edge of the Bunsen flame. Dimming the lights in the laboratory will heighten the effects.

## *Did you know? About firing fireworks*

*Index 01.03*

## *Person profile - Darryl Flemming*

*Index 01.04*

## *Colours in fireworks - demonstration*

*Index 01.05*

Apparatus and equipment
- Bunsen burner
- Heatproof mat
- Pump action spray bottles, *eg* used for cleaning agents, preferably one for each chemical
- Chemicals to test, see list above, about 1 g each
- About 10 cm$^3$ ethanol per solid – **Highly Flammable**
- Eye protection.

What you do

Before the demonstration
1. Make saturated solutions of each of the solids in ethanol where possible. Only very small quantities of the solids are required.
2. Place the solutions in a spray bottle.
3. Label each bottle.
4. Set the nozzles of the bottles to give a fine mist spray, not a jet.
5. The solutions can be kept in the bottles for at least several weeks without deterioration of the plastic or solutions.

The demonstration
1. Darken the room if possible.
2. Set the Bunsen burner to a roaring flame.
3. Spray the solutions into the flames, ensuring these are not directed towards the audience! The flame will turn the characteristic colour of the metal ions in solution.

 Safety
Wear eye protection.

Notes on the experiment

With older students, a hand-held spectroscope could be used to observe emission lines.

It is best to use a trigger action pump spray, not a perfume bottle type squeezing a rubber bulb.

The demonstration is taken from T. Lister, *Classic chemistry demonstrations*, London: Royal Society of Chemistry, 1995.

Results

| Firework chemical | Colour in flame | Notes |
|---|---|---|
| sodium chloride | yellow / orange | Compare to street light colour |
| potassium chloride | lilac | Pink when viewed through blue glass |
| copper(II) carbonate – **Harmful** | green / blue | White flares may also appear |
| strontium chloride | red | |
| calcium chloride – **Irritant** | brick red | Transparent crystals tend to melt and drip easily |
| barium chloride – **Toxic** | apple green | Colour is short-lived |
| lithium chloride – **Irritant** | red | |
| zinc chloride – **Corrosive** | none | |
| magnesium chloride | white / none | |
| iron filings | gold sparks | Can be difficult to get sparks due to temperature |

Answers
Most colours should be the same as in the table. Copper may be different – this appears green in a Bunsen flame. Magnesium metal produces white flares, but ions are typically non-coloured in a Bunsen flame.

## Chemical elements in fireworks

*Index 01.06*

This resource could usefully be used as a follow up activity to **What is in a firework?** or **Colours in fireworks**.

Learning objectives
- To learn flame test colours.
- To find out which chemical elements are used in fireworks.

Time required
About 20 minutes

Answers

1. Answers to the table are given in blue

| Chemical element | Symbol | Metal/Non-metal? | What it does in fireworks |
|---|---|---|---|
| Lithium | Li | Metal | Colouring agent - red |
| Sodium | Na | Metal | Colouring agent - gold and yellow |
| Potassium | K | Metal | Fuel and orange-red sparks, colouring agent lilac |
| Magnesium | Mg | Metal | Fuel and bright white flames |
| Calcium | Ca | Metal | Colour enhancer - helps make colours deeper, colouring agent - brick red |
| Strontium | Sr | Metal | Colouring agent - red (crimson) |
| Barium | Ba | Metal | Colouring agent - green |
| Titanium | Ti | Metal | Fuel and produces silver-white sparks |
| Iron | Fe | Metal | Produces gold coloured sparks |
| Copper | Cu | Metal | Colouring agent - blue/green |
| Zinc | Zn | Metal | Fuel and smoke agent |
| Aluminium | Al | Metal | Fuel and makes yellow-white sparks eg in sparklers |

2. These elements may be combined in a firework to make:
   turquoise  -  barium and copper
   violet  - strontium and copper
   citron (pale yellow) - magnesium and sodium

   Firework makers keep their chemical mixtures as closely guarded secrets. It is therefore not possible to give truly accurate answers.

3. sodium, lithium, magnesium, copper, calcium, potassium.

  ## *Did you know?  About blackpowder*

*Index 01.07*

 ## *Investigating blackpowder*

*Index 01.08*

Learning objectives
• To find out about the principles of an explosion
• To work out what makes fireworks explode.

Time required
About 25 minutes

Demonstration:
**The howling jelly baby / reaction between potassium or sodium chlorate(V) and sugar**

This reaction shows the principles of an explosion. Sugar is reacted with potassium or sodium chlorate(V). The sugar can be granulated or as a 'jelly baby'. Two sets of instructions are given.

## Apparatus and equipment for the jelly baby method

- 6 g pure (reagent grade) potassium or sodium chlorate(V) $KClO_3$ or $NaClO_3$ (Oxidising)
- One jelly baby
- Spatula
- Clean, new pyrex boiling tube
- Bunsen burner
- Heatproof mats
- Safety screens
- Eye protection for audience, face shield for demonstrator
- Clamp stand, boss and clamp
- Tongs.

### What you do

1. Clamp the boiling tube at an angle. Set the safety screen between the apparatus and audience
2. Put the sodium or potassium chlorate(V) in the tube. Sodium chlorate(V) requires less heating, so may be preferable.
3. Heat the sodium chlorate(V) until it is molten. Bubbles of oxygen will be seen as the compound begins to decompose.
4. Turn off the burner. Drop the jelly baby into the tube using tongs. Stand well back.

 **See CLEAPSS School Science Service special risk assessment for further details – available to print from the enclosed CDROM.**

Equations for the reactions are:-

$$2NaClO_3(s) \rightarrow 2NaCl(s) + 3O_2(g)$$

$$C_{11}H_{22}O_{11}(s) + 3O_2(g) \rightarrow 9C(s) + 3CO_2(g) + 11H_2O(g)$$

### Disposal

Soak the tube in water for about 15 minutes, then clean with a brush and detergent. Residues may be washed down the drain.

## Apparatus and equipment for the method using sugar

- 6 g potassium or sodium chlorate(V) $KClO_3$ or $NaClO_3$ (Oxidising)
- 2 g granulated sugar
- Splint
- Metal dish / ceramic tile
- Heatproof mat
- Safety screens
- Eye protection.

### What you do

1. Weigh the masses of sugar and potassium chlorate(V) accurately.
2. Mix the sugar and potassium chlorate together gently by pouring from one piece of paper to another. (Do NOT grind it together in a pestle and mortar)
3. Make a pile of the mixture in the dish or on the tile.
4. Make a small depression in the centre of the pile.
5. Ignite from a distance using a wooden splint attached to a metre rule.
6. Stand back and make sure that the students are behind the safety screen.
7. After a few seconds the pile of reagents will burst into a lilac flame. Flames shoot up to 1 m high. The reaction is over in seconds.

The precise stoichiometry for this reaction is not known. The reaction will not occur if the reagents are not in the correct ratios – **see CLEAPSS School Science Service special risk assessment for further details – available to print from the enclosed CDROM.**

**Safety**
Wear eye protection.

**Disposal**
There is very little residue from this reaction. Any solids remaining can be washed down the drain. If the reaction does not 'go' after 2 minutes, wash the reaction mixture quickly down the sink with copious amounts of water.

**Discussion**
Both reactions produce explosions. Students should observe flames, heat, light and in the case of the granulated sugar reaction, very little solid residue. Discuss with the students what must have happened to the solid materials, working towards the idea that a lot of gas must have been produced. In a firework, the same sort of reaction occurs, with a large volume of gas being produced very quickly, accompanied by a lot of energy which we experience as heat and light. The heat expands the gases very rapidly causing the explosion. This is how all explosive materials work.

In a firework, blackpowder (gunpowder) is the explosive. It contains fuel and a chemical which can supply oxygen.

**Answers**

1. Potassium / sodium chlorate  $KClO_3$ or $NaClO_3$
   Sugar $C_{11}H_{22}O_{11}$
   Sulfuric acid  $H_2SO_4$

2. The sugar is acting as the fuel.

3. The potassium or sodium chlorate supplies oxygen.

4. The word equations are:-

   carbon + oxygen → carbon dioxide

   sulfur + oxygen → sulfur dioxide

   The products are gases at room temperature.

5. The explosion happens because large volumes of gases are produced in a small space, together with a lot of energy being released. This makes the gases move fast and creates a flame. This is what happens in a firework.

## *Extension work*

**Answers**

This activity leads to the calculation of the ratio of components in gunpowder.

1. $4KNO_3(s) + 7C(s) + S(s) \rightarrow 3CO_2(g) + 3CO(g) + 2N_2(g) + K_2CO_3(s) + K_2S(s)$

2. The sulfur dioxide reacts eventually with the potassium to produce potassium sulfide. In fact, the reaction shown is an oversimplification - there is a complicated series of reactions involving the sulfur. See M .S. Russell, *The chemistry of fireworks*, RSC, 2000 for more details.

3. There are eight molecules of gases and two solid products.

4. The gases are produced very quickly with lots of energy in a small space. This causes an explosion.

5. The percentages are: potassium nitrate 77.7%, carbon 16.1% and sulfur 6.2%. The working is:-

$M_r$ value for $KNO_3$ = 101.

Using the mole ratios in the equation gives these masses:-

| | | | |
|---|---|---|---|
| 4 moles $KNO_3$ has a mass of | 4 x 101 | = | 404 g |
| 7 moles C has a mass of | 7 x 12 | = | 84 g |
| 1 mole S has a mass of | 1 x 32 | = | 32 g |
| Total mass | | = | 520 g |

| Percentages: | $KNO_3$ | 404 / 520 x 100 | = | 77.7% |
|---|---|---|---|---|
| | C | 84 / 520 x 100 | = | 16.1% |
| | S | 32 / 520 x 100 | = | 6.2% |

Notes on the calculation

The mixture is close to a traditional gunpowder mixture which has the ratios 75:15:10. This is similar to the mixture used in the Gunpowder Plot. Gunpowder has been made using different ratios - initially in the 13th century the proportions were 37.5: 31.25: 31.25. By the time of Guy Fawkes these were modified to 75:12.5:12.5. Blackpowder for fireworks is made to different compositions depending on the type of firework. A typical mix for a lifting charge or burst charge such as in a mine would have the 75:15:10 ratio.

The reaction of the components in blackpowder is complex, so the stoichiometry cannot be precise. The reaction given here is an approximation for calculation purposes.

This information is available over the internet. However the RSC recommends that students or teachers do not make and/or test gunpowder. The calculation was prepared following advice from CLEAPSS.

*Index 01.09*

## Making blackpowder

### Learning objectives
- To find out about the processes involved in preparing the ingredients of gunpowder
- To apply knowledge about physical processes to a 'real life' process
- To learn about science in a historical context.

### Time required
About 40 minutes

### Notes on the activity

The activity is a comprehension/directed activity related to text (DART) which explores how the ingredients for gunpowder were prepared at the Royal Gunpowder Mills in Waltham Abbey, north of London. The text is fictional and is based on a genuine account of a visit to a gunpowder factory in 1861. The style is written to fit with accounts of Victorian England, such as that of Henry Mayhew, so the language used is deliberately grammatically poor. Try to encourage the students to imagine the black, powdery atmosphere, the smells, the poor, cramped working conditions and to hear the voice of Alice talking with a 'working class' accent. This would be a good piece for reading aloud, if possible.

Answers to the questions focus mainly on the chemistry involved in the processes to prepare and mix the ingredients. A lot more happened to gunpowder in various forms – the Royal Gunpowder Mills had a strong reputation for producing high quality gunpowder and for using the best procedures for manufacture. At its peak about 5000 people worked in the factory situated on the river Lea in Hertfordshire. Further research and cross curricular activities are possible on this topic. Consult the websites for further information.

### Answers

1. Information about charcoal making includes:-
   - alder and willow trees are cut into 3 foot (about 1 m) long sticks
   - sticks are placed in a cylinder
   - cylinders are placed in a furnace
   - the wood burns partially and makes a black stick
   - the charcoal is cooled and ground to a powder.

2. Carbon

3. The carbon would burn to carbon dioxide gas. This is not charcoal – it would be given off in the combustion reaction.

4. Alice says 'half a day' which, if she works 12 hours, means 6 hours.

5. Carbon monoxide gas may escape.

6. The rocks cannot be used as they are.

7. Sulfur dioxide may have caused Ernie's cough. It has a choking smell and is toxic. The gas is formed when sulfur burns in air.

8. For example:-
   'Water is added to the grough and the soluble salts dissolve. Insoluble impurities sink to the bottom. The solution is boiled and the liquid transferred to a cooler. The solution evaporates slowly and crystallises.'

9. The percentages are 75% potassium nitrate, 10% sulfur and 15% charcoal.

10. The mix must be right otherwise the powder will not explode.

11. The turner has teeth to help break down small lumps and to ensure all the ingredients are mixed thoroughly.

12. If Alice works a 12 hour day and she can make 20 charges, this is 12 / 20 = 0.6 hours each charge, or 36 minutes.

13. The mass of one charge of gunpowder is:-
    40 ÷ 22 or 0.454 x 40 = 18.16 kg

14. The safety precautions Alice describes are:-
    - changing clothes  to avoid taking blackpowder to a place where there might be a flame, eg the fireplace at home
    - not smoking in the factory
    - wearing leather shoes – metal soled ones may make a spark which could cause an explosion
    - not going 'between houses' so materials which could cause explosions cannot mix other than in the houses
    - not wearing anything metal – this could create a spark and an explosion.
    Other safety procedures include:-
    - wearing protective clothing, such as safety glasses, gas masks and gloves.
    The measures were not always effective – there were frequent explosions from the higher risk operations such as pressing.

15. Alice's comments on her working conditions include:-
    - a 12 or 14 hour working day
    - poor pay
    - getting hurt or injured
    - being beaten for falling asleep while working
    - being fined for not obeying the rules
    - working as a young child
    - doing heavy labour turning the mixing box
    - lifting heavy bags to weigh the ingredients
    - working with the permanent risk of explosion.

## *Key words*

*Index 01.10*

## *Further information, references and source material*

Websites about Guy Fawkes and the Gunpowder Plot include:-
*http://www.fireworks.co.uk/heritage/index.html#guyfawkes* (accessed November 2003).
*http://www.gunpowder-plot.org/* (accessed November 2003).
*http://www.bbc.co.uk/history/games/gunpowder/index.shtml* (accessed November 2003).

Websites about fireworks include:-

*www.kimboltonfireworks.com* (accessed November 2003).

*homepages.enterprise.net/saxtonsmith/fw/glos2.htm* (accessed November 2003) is a comprehensive glossary of fireworks terminology. There are a few inaccuracies in the chemistry, but overall this is a very useful website.

*www.pbs.org/wgbh/nova/kaboom/anatomy.html* (accessed November 2002) has a very good description of a firework, including how one works, with a good diagram.

There are many websites with information about gunpowder. One giving a description of the processes involved is:
*www.argonet.co.uk/users/cjhicks/gphis.html* (accessed November 2003).

The website for the Royal Gunpowder Mills is:
*www.royalgunpowdermills.com* (accessed November 2003).

The Royal Gunpowder Mills site is now open for visitors. The address is:-
Royal Gunpowder Mills
Beaulieu Drive
Waltham Abbey
Essex EN9 1JY
England

Telephone: + 44 (0) 1992 707 370
Fax: + 44 (0) 1992 707 372
email: *info@royalgunpowdermills.com*

### References

B. Crystall, *New Scientist*, 2001, 172(2315) 39–43.
T. Lister, *Classic Chemistry Demonstrations*, London: Royal Society of Chemistry, 1995.
M.S. Russell, *The chemistry of fireworks*, Cambridge: Royal Society of Chemistry, 2000.

# Crystal chemistry

# RS•C

## Crystal chemistry

**Summary**
These resources explore the influences crystalline substances have in our lives.

| Resource name | Index | Type | Age range | Topic | Media |
|---|---|---|---|---|---|
| What is a crystal? | 02.01 | Experiment | 11–16 | Crystals | |
| Key words | 02.02 | Glossary | 11–16 | Glossary | |
| Did you know? About a girl's best friend – diamonds | 02.03 | Literacy / DART | 11–16 | Diamonds | |
| Big or small? | 02.04 | Experiment | 11–16 | Crystal growth | |
| Growing crystals | 02.05 | Experiment | 11–16 | Crystal growth | |
| Growing snow crystals | 02.06 | Demonstration | 11–16 | Crystals | |
| Did you know? About snow | 02.07 | Literacy / DART | 11–16 | Snow crystals | |
| The body beautiful: protein crystals | 02.08 | Careers and DART | 11–16 | Careers | |
| Did you know? About X-ray crystallography | 02.09 | Information | 14–16 | X-ray crystallography | |
| Liquid crystals | 02.10 | Literacy / DART | 14–16 | Liquid crystals | |
| Ideas and evidence activity | 02.11 | Scientific enquiry | 11–16 | What is science? | |

Key:    Interactive student activity    Photocopiable and printable worksheet    Projectable picture resource

### Issue
Crystals have been known to societies for thousands of years. New crystals are being made all the time from inorganic and organic substances. In what ways are our lives influenced by the crystals around us?

### Chemical topics
- properties of solid substances
- features of crystals
- solubility and saturation
- X-ray crystallography.

### Scientific enquiry issues
- controlling conditions for an experiment
- considering how to test 'hearsay'
- how a scientist does her/his job
- how research can be applied practically.

Finally, in the **Ideas and evidence activity** students are invited to consider an 'alternative' view about crystals – that rather than just being solid, inanimate substances which we use for specific purposes, crystals are considered by some to be 'alive' and capable of holding and transmitting and influencing well-being.

## *What is a crystal?*

*Index 02.01*

### Learning objectives
- Solid substances can be crystalline
- In crystals, particles are arranged in an orderly way.

### Time required
About 30 minutes

This activity begins by encouraging students to present their own ideas about crystals. They then look at a range of solid materials to decide if these are crystalline. From this the characteristics of crystals can be established. The activity supports teaching on states of matter and could be used in this setting.

The first part of the activity could be done as a class brainstorm to collate ideas. This would provide a focus for review at the end of the lesson. Students can make their own records according to the instructions.

It is probably best not to lay out the samples until the first part has been completed, so students can respond without being prompted.

Allow students to circulate around the room with the samples displayed – give time to look at around ten each. With sufficient variety, students will have different results to discuss. Alternatively, if time is short, a static display could be arranged in front of the class and teacher-led discussion could take place with students grouped around this.

In reviewing the results, introduce crystal terminology – see **Key words**. Metals and plastics may be difficult to perceive as crystals; glass may be believed to be crystalline, so these points may need discussion. The key point to reinforce is that a crystal is any substance in which the particles are arranged in a regular lattice. The term 'amorphous' can be introduced to describe non-crystalline substances.

Apparatus and equipment
- Samples of materials – (the list below is intended as a guide only)
- Hand-lens (per group)
- Petri dishes (or similar) to display the samples
- Labels.

*non-crystalline (amorphous)*
rubber (*eg* rubber gloves), jelly cube, paper, plant leaf, wood, woven material, glass, butter, lard, chocolate, sponge, ceramics.

*crystalline*
metals, *eg* paper clip, magnesium ribbon, copper sheet, zinc granules;
rocks, *eg* granite, basalt, calcite, sand; gemstones, *eg* diamond, ruby, emeralds, amethyst (volunteers to donate jewellery); minerals and other compounds: *eg* copper(II) sulfate **(Harmful)**, potassium aluminium sulfate, ice, sodium chloride, sugar, quartz; non-metallic elements, *eg* sulfur, iodine **(Harmful)**, graphite.

*mixed/composite*
soil, clay, carbon fibre sports equipment, coal, plastic bag (low density poly(ethene) has amorphous and crystalline regions).

 Safety
Wear eye protection.

Preparation

Clearly labelled samples of the solids should be prepared in dishes and placed around the room. Best of all would be to use items made from the materials to reinforce their use – examples may include china, liquid crystal display stopwatch, sports equipment (usually composite), computer chip etc.

Have a wide range of materials and ensure that they are not displayed in any particular order. To make the display after the first part of the lesson, give out a range of samples to students working on benches. They can begin by examining these then swap.

Answers

1. Answers may include: regular shape (correct); shiny (not always), transparent (not always), coloured (not always), cannot bend it, cannot see the shape.

   It may be difficult to draw out the answer 'regular shape', as this may appear least convincing, especially for metals. Use of hand-lenses may help – metals can appear crystalline for example, crystals can be seen in zinc coatings on outdoor items like lamp-posts.

2 & 3. See list above. This is touching on some difficult areas - the answer to this question could run to pages. Here are two substances which students may find hard to explain:

   *Glass:* Students may believe that glass is a liquid. This is not true, glasses are amorphous solids. There is a fundamental structural divide between amorphous solids (including glasses) and crystalline solids. Structurally glasses are similar to liquids, but that doesn't mean they are liquid.

It is possible that the 'glass is a liquid' urban legend originated within a misreading of a German paper on glass thermodynamics. For more information see *www.glassnotes.com/WindowPanes.html* (accessed November 2003).

*Plastics:* Plastics have crystalline regions, so strictly are mixed. Even the plastic in a carrier bag when stretched will adopt a regular arrangement at the stretch points. This helps a bag carry relatively heavy weights, because the intermolecular bonding between polymer chains in the crystalline regions requires more energy to break than elsewhere.

4. Crystals always have a regular shape.

5. Crystals have different shapes due to the particles in them. If appropriate, discuss that the particles may be ions (as in sodium chloride), covalently bonded molecules (as in iodine), large molecules (as in graphite and diamond and proteins).

6. Liquid crystals are liquids in which the particles are arranged in a regular way. See **Liquid crystals**. Substances which are gases cannot be crystals at room temperature and pressure, because their particles can't be arranged in an ordered way.

7. Heating crystals will destroy the regular arrangement of the particles. Heating increases the amount of kinetic energy and so the crystal lattice can break down as particles move too much to remain in the ordered arrangement. This forms a liquid. The amount of energy required to turn crystalline substances into liquids can be very great – look up the melting points of some substances which are crystals to confirm this.

## *Key words*

*Index 02.02*

## *Did you know? About a girl's best friend – diamonds*

*Index 02.03*

## Growing crystals

This section includes two class experiments and a demonstration. The second class experiment can be extended to an investigation. The demonstration, **Growing snow crystals**, could be set up alongside either of the two class experiments. Further suggestions are given where these are discussed.

## Big or small?

*Index 02.04*

Learning objectives
- Saturated and supersaturated solutions: preparation and what 'saturated' means.
- Temperature affects crystal size.
- Crystals have regular shapes.

Time required
About 40 minutes

This experiment shows the principle of large and small crystal formation. The experiment is often used to demonstrate the cooling processes in igneous rocks. Here, it is a prelude to growing crystals, showing key principles. The demonstration **Growing snow crystals** could be set up and examined towards the end of the lesson to introduce the idea of saturation and growing crystals.

Apparatus and equipment
**For a class of 30 students**
- 20 microscope slides in hot water / heated to about 60 °C
- 20 microscope slides chilled, *eg* in a freezer
- About 40 cover slips
- About 50 g melted salol (phenyl 2-hydroxybenzoate/phenyl salicylate) (Minimal hazard) split into samples for easy access by several groups of students. The salol can be kept in a water bath at about 50 °C.

**For each pair of students**
- Dropping pipette
- Access to a microscope or hand-lens
- Paper towel
- Stopwatch
- Eye protection.

Safety
Wear eye protection.

Notes on the experiment
- In practice, students will need to have easy access to the salol and microscope, in order to prepare a slide, starting timing and watching for crystals.

- The melted salol can be stored in boiling tubes and could be placed in beakers of hot water on benches for access.

- Microscope slides must be dry when the students use them.

- A third option could be added – using a microscope slide at room temperature. This would give a scale of temperatures for the rate of crystal formation.

- The cold slide will produce crystals most quickly, followed by a slide at room temperature and then finally the hot slide. The precise times will vary depending on local conditions. The melting point of salol is 43 °C.

- Examining the slides under a microscope is valuable – seeing crystals appear before your eyes is very exciting. The crystals produced on the hot slide will be the largest and those on the coolest slide will be smallest.

Answers

1. The hottest slide will produce the largest crystals. This is because the rate of cooling was slowest, so more particles were able to join the regular arrangements.

2. Particles form a regular array. They line up alongside other particles, forming bonds between them. This is what makes the salol form crystals.

3. Basalt is formed by quick cooling and has small crystals. Granite is formed by slow cooling and has much larger crystals than basalt.

## *Growing crystals: experiment and investigation*

*Index 02.05*

Learning objectives
- Saturated and supersaturated solutions: preparation and what 'saturated' means.
- Temperature affects crystal size.
- Crystals have regular shapes.

Time required
Stage 1: At least 40 minutes.
Stage 2: At least 30 minutes, two days after stage 1.
Allow time in subsequent lessons to look at progress.
Allow about 30 minutes to 'close' the growing process, including drawing the crystal and writing a conclusion.

The class experiment provides a basic method for making 'potash alum' crystals. The method is based on the 'recipe' in chapter 12 of *'Crystals and crystal growing'* by Holden and Morrison (see references). Students will need to be patient and accurate to get good results. The experiment can be extended by using other substances and/or by investigating answers to the questions suggested.

The demonstration **Growing snow crystals** could be set up and left while students carry out the first or second stage of this experiment. The demonstration indicates the conditions required to produce snow crystals, showing that air can become supersaturated with water.

Requirements per group of students

### For stage 1
- 250 cm$^3$ beaker
- Access to a balance
- Glass rod
- 100 cm$^3$ hot water (50–60 °C – this could be from a kettle, cooled to this temperature)
- Eye protection
- Bunsen burner and heatproof mat
- Tripod and gauze
- 20 g alum (aluminium potassium sulfate, $AlK(SO_4)_2.12H_2O$), plus a few extra grains (students can measure the amount accurately, or this can be pre-measured to save time)
- Watch glass
- Clingfilm to seal the beaker, or a screw top container to store the solution
- Spatula
- 100 cm$^3$ measuring cylinder.

### For stage 2
- Thin wire bent into a 'cobra' shape, see diagram (students could make this if time allows)
- Hand-lens
- Sewing thread (allow about 20 cm)
- Cloth and rubber band to cover the beaker (this needs to be secure; cloth needs to be porous, so not clingfilm or other plastic sheet)
- 250 cm$^3$ beaker
- Access to tweezers
- Bunsen burner and heatproof mat
- Tripod and gauze.

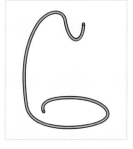

*'Cobra' shape of wire for crystalisation experiment*

Safety
Wear eye protection.

Notes on the experiment

- Time can be saved in stage 1 by having the water and alum pre-measured and hot water available. However, students would miss the opportunities to measure accurately in this case.

- Any good seed crystals can be saved for future experiments, so encourage students to pick several.

- 'Potash alum' crystals are among the easiest to grow, so there should be a good crop of crystals within a couple of days.

- Here is a guide to the mass values and temperatures needed to make saturated solutions for some of the other substances listed in the students' notes:

| Substance | Mass/g/100 cm$^3$ H$_2$O | Temperature/°C |
|---|---|---|
| Sodium chloride | 40.0 | 50 |
| Sucrose | 248.8 | 45 |
| Potassium sodium tartrate, 'Rochelle salt' | 130.0 | 50 |

| Substance | Mass/g/100 cm$^3$ H$_2$O | Temperature/°C |
|---|---|---|
| Hydrated magnesium sulfate, 'Epsom salts' | 31.3 | 40 |
| Disodium tetraborate, 'Borax' | 10.0 | 100 |
| Copper(II) sulfate – **Harmful** | 30.0 | 50 |

- 'Borax' is very cheap and readily available as a cleaning agent.

Other crystal experiments

### Imitation 'geode'
Pour some saturated solution into an empty half walnut shell. Support the shell to prevent it from tipping. Allow the solution to evaporate. This will form an imitation 'geode'. A geode is a piece of rock inside which crystals form.

### Borax 'snowflake'
Borax crystallises more rapidly than most substances, giving white crystals. To make a snowflake, shape three pipe cleaners into a six-pointed star and suspend this into a borax solution. Overnight, the star will become coated with borax crystals.

### Answers

1. A saturated solution is at the limit of solubility – no more solid can dissolve in it.

2. A supersaturated solution holds more than the limit. This sounds strange, but can be demonstrated by adding alum to an unsaturated solution (the added salt will dissolve), a saturated solution (the extra will stay unchanged) and a supersaturated solution (extra alum will stick to the added grains). The point is that 'saturated' has a particular meaning for an amount of substance dissolved in a solvent at a given temperature. It is hard to get a perfect saturated solution. A solution can become saturated from being under-saturated, or supersaturated.

3. When a solution is super-saturated, a crystal added to this will get bigger as the solution reaches saturation.

4. As the solution cools water evaporates from the surface. This increases the concentration of alum in the solution. The solution become supersaturated. The excess is deposited on the crystal, which grows bigger. At the same time, more water evaporates and the growing process continues as the solution deposits excess to reach saturation. The combination of evaporation and deposition keeps the crystal growing.

5. Dirty apparatus may cause crystals to form in unwanted places. To get one single large crystal, there must be no dust grains or other particles around which the alum could crystallise. This is also why using a cobra is preferable to hanging the crystal just from a thread – the thread at the surface can act as a nucleus for crystals to form, affecting the size of the main crystal.

## *Growing snow crystals : demonstration*

*Index 02.06*

Learning objectives
- Conditions for growing snow crystals
- Modelling nature in the laboratory
- Formation of snow.

Time required
About 20 minutes to set up before a lesson – preferably well in advance.
Observation takes about 5 minutes, but can be repeated.
The demonstration can be left for several hours, or a school day.

The demonstration is very pleasing, but don't expect an avalanche! The diagram below shows how to set up the equipment. In testing this, a smooth-sided bottle works best. Crystals grew within about 20 minutes.

Snow crystals are grown using the same principles as for other substances – creation of a supersaturated environment at an appropriate temperature, with a nucleus around which crystallisation can take place. Here, in the growth chamber (the plastic bottle) the air is supersaturated with water vapour at a temperature well below 0 °C. These are the conditions required to form snow in the atmosphere. In this case, the thread provides the nucleation site for growing the crystals.

Note that snow is not simply 'frozen water'. In weather terms, this would be hailstones. Snow is a crystalline form of water formed under these specific conditions. **Did you know? About snow** gives more information.

The demonstration could be extended into an investigation to explore more thoroughly the conditions surrounding the formation of snow. This could be aligned with a study of weather patterns. Websites relating to these aspects are given in the reference section.

## What you need

- 500 cm$^3$ smooth-sided bottle with top
- Sponge cut to fit the base of the bottle, about 1.5 cm deep
- About 15 cm nylon thread
- Small paper clip
- Sewing needle
- 4 pins
- Hammer
- About 2 dm$^3$ (litres) dry ice, *ie* solid carbon dioxide
- Stanley knife or sharp scissors
- Sticky tape
- Insulated containers to fit the bottle  - see diagram and text
- Hand-lens.

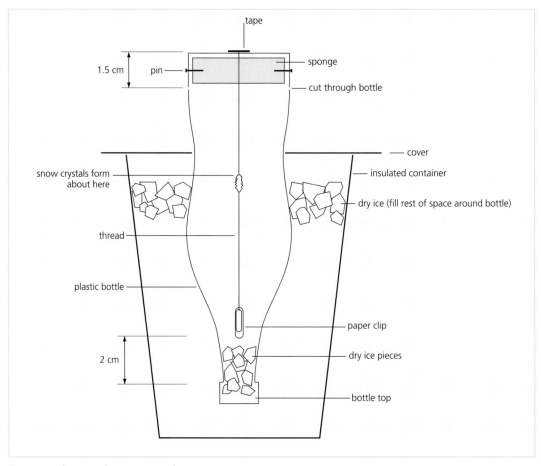

*Demonstration: growing snow crystals*

### What you do

1. Prepare the bottle – cut the base from the bottle about 1.5 cm deep. Cut a sponge to fit neatly inside the base. This does not need to fit exactly. Make a hole near the centre of the base (depending on the bottle, the centre itself can be very hard and dense, as this is the point where the plastic bottle moulding process was completed). This can be tough – use the hammer to make a fine hole with one of the pins. Make four holes around the side of the base.
2. Place the sponge into the base. Secure the sponge by pushing pins through the four holes in the side. Thread the needle with the nylon and push through the hole in the centre of the base so that the thread comes out in the centre of the sponge and can hang freely.
3. Stick the end of the thread on to the outside of the bottle base using tape. Tie the paper clip to the free end of thread. Adjust the length of the thread so the paper clip dangles freely inside the bottle when the base is replaced. To do this, release the tape holding the end of the thread and pull it up the bottle until the paper clip is about 2 cm above the bottle top.
4. Stand the bottle in the container. Place 3–4 pieces of dry ice into the bottle. Fill the gap between the bottle and container with dry ice, packing it tightly. The ice should come about 2/3 up the bottle. Cover the exposed dry ice with an insulating material.
5. Wet the sponge with water. Replace the base of the bottle neatly on to the bottle, trying to ensure a tight fit. Cover the whole set up with more insulating material. Leave the apparatus to stand for about 20 minutes.
6. After 20 minutes, try to look into the bottle. If needed, gently lift the base away from the bottle. Snow crystals should have formed on the nylon thread. These will melt quickly when taken into warm air.

### Notes on setting up the demonstration

- The air inside the bottle must be supersaturated to produce the crystals. It is important that the base sits snugly on to the bottle to help this.

29

- A larger bottle, *eg* a 2 dm³ (litre) clear-sided bottle may be easier to use – but in this case ensure the outside container is sufficiently large and that the amount of dry ice is at least doubled.

- If a larger bottle is used, it may be easier to see the crystals without lifting the base. This will enable them to be photographed.

*Apparatus set up*

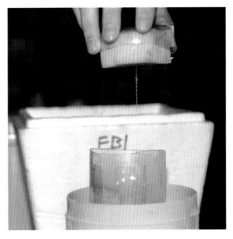

*Arrangement of sponge, thread and clip*

*Snow crystals growing on the thread*

- Dry ice (solid carbon dioxide) can be obtained from university chemistry departments or chemical companies. They may make a small charge for this. Handle with care with tongs or insulating gloves as it can cause frostbite. Use in a well ventilated room.

- A temperature check inside the bottle should show about –15 °C.

- The crystals can be examined with a hand-lens through the bottle.

- The dry ice will last for a few hours. The demonstration could be prepared in the morning and be available throughout a whole school day.

### Answers

1. The snow crystals produced are usually about 5 mm long.

2. The crystals melt quickly when they are lifted out of the bottle. This is because the air around them is no longer saturated and is at a much higher temperature, so the water in the snow changes into water vapour.

3. Snow needs supersaturated air and cold temperatures to form. These conditions are similar to those needed by any other substance to crystallise.

## *Did you know? About snow*

*Index 02.07*

 ## *The body beautiful: protein crystals*

*Index 02.08 (4 pages)*

---

### Learning objectives

- Proteins can form crystals
- Protein crystals are important for finding out how the body works
- Crystal structure can be determined by a technique called X-ray crystallography
- Protein crystals need the same conditions to grow as other substances.

---

### Time required
About 40 minutes

This activity describes the work of a real-life protein scientist who crystallises proteins. This is an exciting task, as protein molecules are much larger and more complex than the average inorganic substance. Although the conditions for protein growth are very similar to those for other substances, the small size of the crystals and complexities of structure mean that specific combinations of reagents must be found to create the optimum conditions for each individual molecule. Part of Julia's work is to find these conditions.

Protein crystals are used in X-ray crystallography experiments. This activity introduces this technique through **Did you know? About X-ray crystallography**. The technique establishes the arrangement of atoms in complex molecules. Knowing the structure is a prelude to finding out how the molecule works in the body. Information about protein structure is passed to drug companies and other researchers to help in the long-term provide better care for diseases and illnesses.

### Answers

1. Both require supersaturated conditions and evaporation of water to make crystals.

2. Protein crystals are grown in a drop hanging from the top of a small container. In **Growing crystals** crystals are grown from a cobra inside a solution. The size is also very different – tiny amounts of solutions containing the protein are needed, whereas for salt crystals much larger quantities of liquids are needed.

3. The sizes and shapes vary because the protein molecules are different.

4. Julia would test which salts and PEGs make the protein crystal grow best. She may also change the temperature and humidity.

5. Proteins make good crystals because their molecules have specific shapes and can be packed into a regular lattice.

6. Protein crystals are needed in science because a lot of scientific research depends on knowing how the body works. Proteins are involved in this.

## *Did you know? About X-ray crystallography*

*Index 02.09*

## *Liquid crystals*

*Index 02.10*

Learning objectives
- To understand a piece of scientific research
- To appreciate the practical applications of scientific research
- To realise that scientific research can be based on good ideas and problem-solving.

Time required
About 25 minutes

# Crystal chemistry

The text describes a research project to develop liquid crystals which can change colour. The text has been simplified to aid students' understanding. The basic principle is that layers of liquid crystal molecules can be stuck together into a flake. The layers can be made to rotate. Rotating the layers in the flake causes the flake overall to reflect light differently. This can be seen as a colour change.

The project aims to develop the flakes into practically applicable formats. Then it would be possible to have paints, paper and computer screens which would change colour when a small electric current passes through them.

The liquid crystals are:
4'-n-pentyl-4-cyanobiphenyl (also known as 5 CB) and PAA-polyarylamide.

Answers

1. These liquid crystals can change colour. The ones in calculators stay black or grey. Also, these are made from stacks of molecules which can flip and rotate. Normal liquid crystals are not in stacks.

2. The flakes can only be made between two glass plates at the moment, which is not very useful. Also, the researchers need to control the flipping movements.

3. The three possible uses are: in paints, paper and on a portable computer screen. There may be a range of different suggestions – here are some:
   - in a monitoring system of a patient who is seriously ill – a monitor could change colour if a current supplying a piece of equipment fails
   - colour change clothing could help emergency workers, record body temperature
   - calculator and mobile telephone screens could be made more exciting and interactive through using colour.

## *Ideas and evidence activity*

Learning objectives
- To consider the place of science as an influence in society
- To consider scientific evidence as a key to understanding false information
- To debate an issue based on hearsay evidence.

This activity centres on the incorrect but common thinking that crystals have 'special powers'. A great deal of readily available false information is circulated proposing crystals as 'cures' or 'aids' to a wide variety of physical and emotional conditions, none of which has any basis in scientific fact. 'Crystal therapists' are available who, for a large fee, will wave a crystal over the body to create an 'aura' to cure stress or whatever ill has befallen the client. Crystals have been used to help search for the Loch

# Crystal chemistry

Ness Monster, as divining tools and in a variety of other ways. This thinking is almost certainly rooted in prehistoric traditions and cultures, continuing today in a variety of pseudo-scientific formats.

Here students have the chance to look at claims made by people who 'believe' in crystals and to challenge these from a scientific perspective.

## Time required
About 30 minutes

## Answers

1. Many people do not think scientifically; people search for easy solutions to problems; there is a historical tradition of crystals as an 'alternative' therapy; people can be persuaded by listening to anecdotes of people 'cured' by crystals; scientific-sounding claims can be persuasive.

2. The claims are not made on scientific evidence.

3. Incorrect statements include:
   • 'Quartz is a powerful energy source'
   • 'Rose quartz... adds positive energy'
   • '...crystals can store and amplify power'
   • 'Red, yellow and orange crystals are... energy-producing'
   • '...cleanse [a crystal] to make sure any stored energy is removed'.

The crystal experiments also have no scientific basis in fact.

## *Further information, references and source material*

### References

A. Holden and P. Morrison, *Crystals and crystal growing,* London: MIT Press 1982, 13th printing 2001.
This very useful book has many 'recipes' for growing crystals

D. E. Sands, *Introduction to crystallography,* New York: Dover Publications Inc, 1993.

### Crystal growing sets
There are several good crystal growing sets available. Contacts include:

Natural Science Industries, UK
3 Tresham Road
Orton
Southgate
Peterborough PE2 6SG

*www.snowcrystals.net* (accessed November 2003) and *www.snowflakebentley.com* (accessed November 2003) both contain excellent images of snow crystals.

Within this chapter there are a number of images taken from the Amethyst Galleries' Mineral Gallery at *mineral.galleries.com* (accessed November 2003). Any reproduction of these images must follow the restrictions outlined in the galleries' copyright notice below:

### Amethyst Galleries' copyright notice
Unless otherwise noted, all mineral descriptions and images, plus the related descriptions on this server are the property of Amethyst Galleries Inc., and may not be copied for commercial purposes. Permission to copy descriptions and images is granted for personal and educational use only. All such copies must include this copyright notice and explicit references to the URL *mineral.galleries.com* (accessed November 2003).

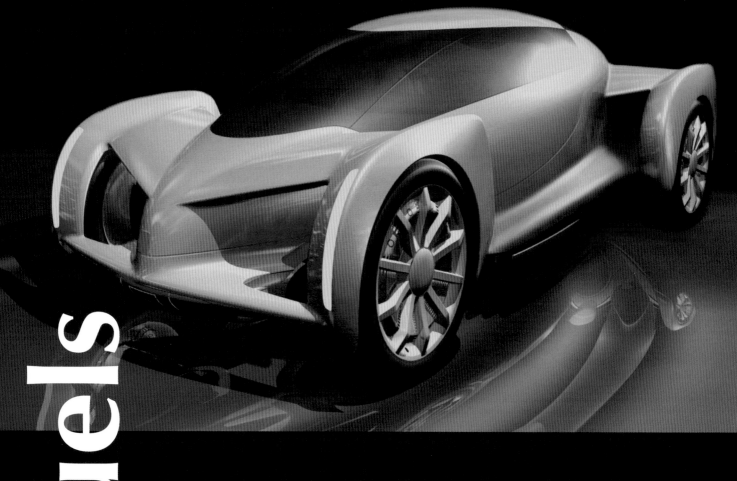

# Future fuels

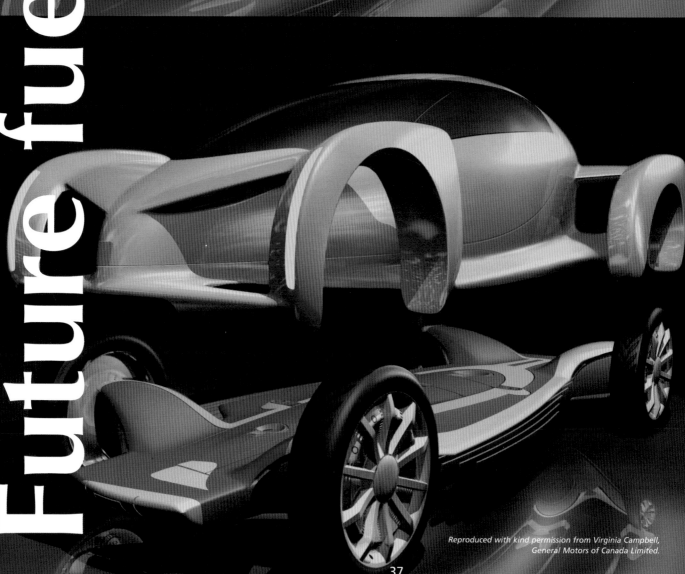

*Reproduced with kind permission from Virginia Campbell,
General Motors of Canada Limited.*

# RS•C    Future fuels

**Summary**
This collection of resources shows that hydrogen fuel cells are being considered as a way of reducing carbon emissions from vehicles.

| Resource name | Index | Type | Age range | Topic | Media |
|---|---|---|---|---|---|
| Collision course – which car? | 03.01 | Paper based exercise, literacy / DART | 11–16 | Alternative fuels | P |
| Did you know? About the greenhouse effect | 03.02 | Literacy / DART | 11–16 | Greenhouse effect | ◑ P |
| Key words | 03.03 | Glossary | 11–16 | Glossary | ◑ P |
| Debate: Collision course – which car? | 03.04 | Role play, analysis, discussion, debate and report writing | 11–16 | Alternative fuels | P |
| Fuel cell vehicles | 03.05 | Paper based exercise, literacy / DART | 11–16 | Fuel cells and batteries | P |
| Fuelling the fuel cell | 03.06 | Modelling exercise or Interactive activity | 14–16 | Modelling formation of water | ◑ P |
| Demonstration: the hydrogen – oxygen rocket | 03.07 | Demonstration and questions | 14–16 | Energy release from combustion | P |
| Demonstration: Bubble trouble | 03.08 | Demonstration and questions | 14–16 | Energy release from combustion | P |
| How much energy? | 03.09 | Numerical problems | 14–16 | Bond breaking, bond making and chemical equations | P |
| Splitting water – making hydrogen and oxygen | 03.10 | Demonstration and questions | 14–16 | Electrolysis of water | P |
| Splitting water – making hydrogen and oxygen | 03.11 | Experiment | 14–16 | Electrolysis of water | P |
| Hydrogen fuel cells | 03.12 | Literacy / DART | 14–16 | Hydrogen fuel cells | P |
| Fuel cell cars: The GM AUTOnomy | 03.13 | Literacy / DART | 14–16 | New technologies | P |

Key:  ◑ Interactive student activity    P Photocopiable and printable worksheet    ◎ Projectable picture resource

# Future fuels

### Issue

Governments in the western world have challenged car manufacturers to reduce carbon emissions from cars, partly because of global warming and because petrol is a non-renewable resource. Hydrogen powered and hybrid cars are being considered. Can hydrogen fuel cells help and what might future cars look like?

### Chemical topics

- Fossil fuels
- Electrolysis of water
- Energy release from combustion
- Chemical equations
- Changes to the amount of carbon dioxide in the atmosphere.

### Scientific enquiry issues

- Scientific developments are affected by pressure from governments
- How science can address environmental questions
- Evaluation of evidence to support a conclusion
- Reporting science to the public.

### Notes on using these activities

The activities are all free-standing so may be used independently of each other. Some activities are paired together as alternatives addressing the same issue or aspect. These are indicated in the text below.

**Collision course – which car?** sets the scene for the other activities, but could be used to support existing material on alternative fuels in its own right. The activity involves using a report from 'The Times' newspaper to consider the issue of replacing petrol driven cars with other fuels. The text is presented alongside questions so students do not have to hunt for the answers, some prior knowledge is required.

**Debate: Collision course – which car?** is a role play. This activity could be used as an alternative to **Collision course – which car?**, as the issue addressed is the same. However, the role play would probably work best if students were introduced to the topic through an activity like the report. The role play takes the form of a government meeting organised to discuss and decide upon which type of car should be recommended for Britain in the future. The role play mimics the information in the report, but extends this to include as many members of a class as possible. Written work from the activity should be in a format which is accessible to the 'general public'.

**Fuel cell vehicles** begins by describing a fuel cell and shows that water is needed. This could be nicely followed by **Fuelling the fuel cell** which uses models to find out about water. An alternative interactive activity is also available on the CDROM. Two demonstration reactions; **The hydrogen-oxygen rocket** and **Bubble Trouble** are given to show how hydrogen and oxygen can be combined to make water. Higher level students can calculate the amount of energy generated in **How much energy?**. **Splitting water – making hydrogen and oxygen** is also shown in two ways – either as a demonstration using the Hoffmann voltameter (with a novel twist) or as a class experiment.

In **Hydrogen fuel cells** the theme is drawn together and students look at how a hydrogen fuel cell works in practice. This could be done either by completing the paper-based activity including photographs, or by demonstration if a hydrogen fuel cell is available. In **Fuel cell cars: The GM AUTOnomy,** the latest car technology (in 2002) is presented to show how the science may apply in practice to fuel the future.

## Report: Collision course - which car?

*Index 03.01*

### Learning objectives
- Burning fossil fuels contributes significantly to the greenhouse effect
- Low carbon technology (LCT) is being developed to reduce carbon emissions
- LCT means burning less carbon-based fuels
- Hydrogen is an alternative fuel
- Hydrogen-powered and hybrid cars are being considered instead of petrol-driven cars
- Governments need to use science to help make decisions.

### Time required
About 40 minutes

This activity uses a report from *The Times* newspaper to show that the government is considering how to lower Britain's contribution to global carbon dioxide emissions. Since the report was published, the document 'Powering Cleaner Vehicles' has been made available by the Department for Transport, which discusses the strategy in detail. To access the document see page 60.

Note - some questions require knowledge not directly in the text, so the activity could be used to support a lesson on alternative fuels using existing material. The answers are presented together with the questions to facilitate checking. The activity could also be used to teach note taking and how to find key words in a text. It could be concluded usefully by a discussion of how governments rely on science for making decisions. Also, this activity provides background for the role play **Debate: Collision course - which car?** and may be used as preparation for this.

The activity could also be used as a stand-alone piece of homework, however, teachers should be aware of relatively high reading age of the passage.

### Minister is set for collision on move to hydrogen cars - Answers

| Questions | Answers |
| --- | --- |
| What is meant by fossil fuels? | Fuel made from decayed animals and plants which lived thousands of years ago, *eg* coal, oil and natural gas. |
| Why do people think these fuels should be 'phased out'? | Burning them contributes to air pollution and the greenhouse effect. |
| What does the government mean by 'more efficient' petrol or diesel cars? | Cars which use less fuel per mile, or produce less pollutants per litre of fuel burned. |
| Why is hydrogen called an 'emission-free' fuel? | It does not produce any polluting gases. |
| Why is it important to reduce carbon dioxide production? | Carbon dioxide is one of the main greenhouse gases. Reducing emissions will reduce the greenhouse effect. |
| What is 'low-carbon technology'? | Engines which use less carbon-based fuel and/or produce less carbon in exhaust gases. |

How would using hydrogen-fuelled cars help reduce carbon emissions?

Hydrogen-fuelled cars do not produce any carbon-containing compounds in the exhaust gases.

What is a 'hybrid' vehicle?

A vehicle which can run on petrol or an electric motor.

What does the Minister mean by a 'renewable' source of or hydrogen?

A source which does not deplete the Earth's resources of fuel An example is using solar energy to electrolyse water.

Why does the Minister think having a renewable source of the gas is needed?

Hydrogen cars would be of no benefit if they use up fuel/energy simply to make the hydrogen needed to power the car. If a 'renewable' source is used, then the benefits are very good, see above.

What is meant by 'fuel efficiency'? Why is this important?

This is the distance a car may travel on one litre of fuel. The further the distance, the more efficient the car, and the less pollution per mile produced.

Why does the BMW chairman disagree with the Minister's views about hybrid cars? Give two reasons.

He thinks that hybrid engines only 'shift the burden of emissions'. This means that we will still get the emissions, but not from cars. To charge up an electric motor will require energy and that will have come from burning a fuel. Also he wants to sell the BMW hydrogen car.

Why does the car industry expert disagree with the Minister's views about hybrid cars?

Two reasons: using hybrid cars will slow the move to non-polluting hydrogen and they are more expensive than a vehicle with one engine.

Why is it useful to have a car which can run on hydrogen or petrol?

A driver could still buy petrol for the car while waiting for hydrogen to become available in garages.

  ## *Did you know? About the greenhouse effect*

*Index 03.02*

  ## *Key words*

*Index 03.03*

## Debate: Collision course - which car?

*Index 03.04*

---

### Learning objectives
- Burning fossil fuels contributes significantly to the greenhouse effect
- Low carbon technology (LCT) is being developed to reduce carbon emissions
- LCT means burning less carbon-based fuels
- Hydrogen is an alternative fuel
- Hydrogen-powered and hybrid cars are being considered instead of petrol-driven cars
- Governments need to use science to help make decisions
- Writing a report about science for the public.

---

### Time required
At least 60-90 minutes student preparation time (more may be required).
40-60 minutes to run the meeting.
Time afterwards to make a display of prepared materials.

The learning objectives for this activity are identical to those for the newspaper report based activity **Collision course - which car?** with the addition of a written report, so the activities can either be used separately or together. **Collision course - which car?** provides useful background and scene-setting for this role play debate activity. The activity could be used in the context of a unit on alternative fuels and energy.

A role play is always a very exciting activity. The level of preparation required is high, but once done, the lesson runs itself. Students are often surprisingly responsible when allowed to run their own lesson once in a while – and individuals taking serious roles can often bring out hidden talents and skills which traditional activities miss. Do try it – some suggested advice given below may help.

### Advice about running the role play

Roles can be allocated to individuals or small teams of a maximum of three students. Any more than this and there is not enough work to go round. Ensure that the Transport Minister is someone who will 'play fair' – this may not be the most obvious person, as sometimes even the most difficult student can respond when given this type of role. Alternatives for the Ministerial role allow the opportunity for added creativity. To add an extra dimension, invite someone from outside. This could be an older student (*eg* a post-16 student studying chemistry), a colleague (*eg* another member of the Science Department, senior manager or the Headteacher, school governor or Chair of Governors), or a friendly local politician (perhaps the Member of Parliament or local government councillor responsible for transport). If the role play is run with a significant local figure as the chair, then get the local paper along to generate positive publicity for the school.

A safe option is take on the role yourself, but the disadvantage is that this places you as arbiter, rather than allowing the students to present to an outsider or one of their own. Another alternative may be to appoint a 'Ministerial Team', so the decision is not made by one student alone – but ultimately having one person in charge of the entire meeting is recommended.

### Whole class involvement

The material presented at the meeting is best prepared by teams of students working together within the roles suggested. The teams can choose a representative to make the presentation. Teachers could suggest that students acting as representatives of car makers, low carbon technology, fuel suppliers, and the car industry, prepare Press Releases before the meeting for briefing the journalists, and that the journalists write reports after the meeting for their respective magazines or newspapers. This will give a good spread of different types of report.

The reporters will need to work out questions most appropriate for their readers' concerns. Identify in advance the papers and magazines represented. Examples might include the main daily tabloid and broadsheet newspapers, *New Scientist* and *What Car?*. Teachers could ask students for their own suggestions for magazines.

A reporting team is sent to the meeting from each newspaper or magazine, with the teams meeting beforehand to research and prepare questions, and afterwards to prepare their articles based on the outcome.

### Preparation time

Students will need plenty of preparation time for this exercise. Most students enjoy taking role plays seriously, but they do not want to be 'shown up' and complaints arise when they feel 'cheated' of time to prepare. Give plenty of notice of the meeting date, place and time. This point is especially important if an outside guest is invited to chair the meeting.

### Running the role play lesson

A full lesson will be required for this, normally 40–60 minutes. Points to remember:
- Arrange the furniture appropriately
- Take the meeting seriously and expect the students to do the same
- Restrict the time for presentations so everyone has a fair chance. A maximum of 5–10 minutes should be plenty
- Ensure the Minister keeps time fairly
- Ask the Minister to prepare a running order in advance so everyone knows when their turn is
- Allow time for questions and for the decision to be made after the presentations.

### A record of the event

Teachers might like to consider making a visual record either as a video or photographs for use later. A display of the press releases and articles would also be a valuable conclusion.

### Debriefing

It is recommended that teachers allow time to ensure the students are debriefed afterwards. To do this, state explicitly that students are no longer in their roles *eg* by specifically saying *'X is not the YY'* *etc*, and that the discussion is confined to the meeting alone and has ended.

## *Fuel cells*

There are three parts to this section. In the first, students find out what a fuel cell is. Next, they see water made by reacting hydrogen and oxygen and in the third they split water using electrolysis.

Learning objectives
- What a fuel cell is and how it works
- Water molecules can be made by reacting hydrogen and oxygen gases
- Water molecules can be split by electrolysis
- Energy is given out when water is formed
- Tests for hydrogen gas and oxygen gas.

## *Fuel cell vehicles*

*Index 03.05*

### Time required
About 30 minutes

This activity is a short introduction to fuel cells, making the contrast between a fuel cell and a battery.

### Answers

1. Features of a battery – creates electricity, uses solid materials, will run out.
   Features of a fuel cell – creates electricity, uses gases, will not run out if gas supply is maintained.

2. The fuel cell. Batteries create a stockpile of used chemicals which need to be disposed of. The fuel cell can keep going indefinitely provided the membrane is not poisoned and a supply of gas is maintained. Therefore less of the Earth's resources are needed to make fuel cells and they create less waste.

3. (a) A battery will always run out   (b) A fuel cell needs a constant supply of gas.

4. Compounds containing hydrogen which students may think of are: water, hydrocarbons like methane, alcohols like ethanol and methanol, carbohydrates, sugars. These would have to be split to release the hydrogen gas.

5. There was no perceived need. Batteries were cheap, easy to make and easy to use. Other technologies were developed which were taken up much faster, perhaps because the scientists or engineers involved had more money, were better known, or were more self-promoting than Sir William Grove. Also, the potential for fuel cell technology probably could not be foreseen at that time.

## *The fuel cell reaction*

The following activities show that water can be made in a chemical reaction and split using electrolysis.

Teachers are advised to look at all the options and choose the most appropriate combination for their classes.

There are a number of different activities in this section that teachers can choose from:
- Molecular modelling exercise: **Fuelling the fuel cell**
- **Demonstration: The hydrogen – oxygen rocket** – this involves making water by exploding hydrogen and oxygen gases in a rocket
- **Demonstration: Bubble trouble** – this involves splitting water by electrolysis and exploding a mixture of the gases bubbled through soap solution

# RS•C    Future fuels

- **How much energy?** A short paper-based calculation of bond energies. The molecular modelling exercise **Fuelling the fuel cell** should be completed before attempting this activity
- **Demonstration: Splitting water** – making hydrogen and oxygen – two different techniques are given; one involves electrolysis using the Hoffman apparatus and the other uses standard electrolysis equipment and an OHP
- **Class experiment: Splitting water – making hydrogen and oxygen** – this involves splitting water using electrolysis.

## *Fuelling the fuel cell*

*Index 03.06*

**Time required**
About 10–20 minutes

This activity provides an introduction to the experiments/demonstrations. Allowing students the opportunity to make molecular models and to model the reaction is useful, as they will develop a stronger feel for the molecules involved and be able to understand the written equation more easily.

One type of molecular modelling kit is available from Spiring Molymod at *+ 44 (0)1403 782387 molymod@globalnet.co.uk*, **www.molymod.com**
Other types can be purchased through most school equipment catalogues. See also **www.chemsoc.org/learnnet**, and look in the links area.

### Answers

1. The equation for the reaction between hydrogen and oxygen gases is:
   $2H_2(g) + O_2(g) \rightarrow 2H_2O(l)$

2. A complete sentence might be:
   Two molecules of hydrogen gas and one molecules of oxygen gas break up into atoms and recombine to make two water molecules and releasing energy.

3. The water molecules must break up and the atoms recombine to make hydrogen and oxygen gas molecules. This requires energy.

The students could also use the interactive activity which allows them to 'drag and drop' atoms and bonds to make models of molecules.

### Demonstration: The hydrogen – oxygen rocket

*Index 03.07*

This is a very satisfying demonstration and students and teachers enjoy it.
It is safe providing safety advice is followed.

Apparatus and equipment (including equipment to make the gases if gas cylinders are not available)
- 2 x 500 cm$^3$ empty plastic (PET) bottles *eg* for mineral water, with screw caps
- 2 x 250 cm$^3$ conical flasks with side-arms and rubber tubing
- 2 one-holed bungs to fit the flasks
- Dropping funnels
- Trough
- 2 each of clamps, bosses, stands
- Tongs
- 2 spatulas
- 10 g zinc powder (**Flammable**)
- 5 g manganese(IV) oxide (**Harmful**)
- 200 cm$^3$ 20 volume hydrogen peroxide (**Irritant**)
- 100 cm$^3$ 5 mol dm$^{-3}$ hydrochloric acid (**Irritant**)
- Few crystals copper(II) sulfate (catalyst for producing hydrogen) (**Harmful**)
- Bunsen burner and heatproof mat
- Splint
- Eye protection.

Preparation
Make a waterproof mark about two-thirds up each bottle. Fill both bottles with hydrogen/oxygen mixtures. Prepare the gases as follows:

Hydrogen:
1. Put a few spatulas of zinc into a conical flask. Add a few small crystals of copper(II) sulfate.
2. Fit a dropping funnel through the bung.
3. Secure the flask with a clamp and stand.
4. Place hydrochloric acid in the funnel.
5. Fill a plastic bottle with water.
6. Fill the trough about two-thirds full with water.
7. Place the bottle upside down in the trough to collect the gas by displacement.
8. Open the tap on the funnel allowing acid to drip onto the zinc. Allow air to bubble out for a few minutes.
9. Place the rubber tubing into the bottle and displace the water up to the mark.

Oxygen:
1. Put several spatulas of manganese(IV) oxide in the second flask.
2. Fit the funnel and secure in the same way as for the hydrogen preparation.
3. Place hydrogen peroxide in the funnel.
4. Drip the hydrogen peroxide onto the manganese(IV) oxide.
5. Allow air to be displaced before collecting any gas.

6. Place the rubber tubing into the bottle and displace the remaining water.
7. Stopper the bottle using the screw cap.

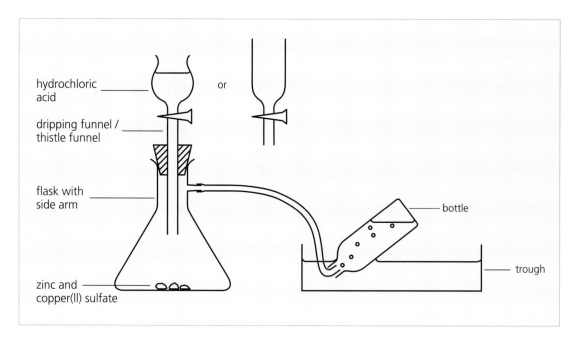

### Disposal
- **Zinc:** Add excess hydrochloric acid to react completely with the zinc. Neutralise and pour down the sink. Alternatively allow the reaction to stop, decant to remove the zinc and pour the zinc chloride solution down the sink.
- **Oxygen:** Filter to remove manganese(IV) oxide and recycle. Wash remaining hydrogen peroxide solution down the sink.

### Launching the hydrogen/oxygen rocket

### Safety points
- Ensure that the audience are behind the 'rocket'
- There is a risk of burns to the igniter – follow instruction 5 carefully to minimise this
- Note that the bang can be loud and ear protection could be used. Do not fire the rocket in an enclosed space.

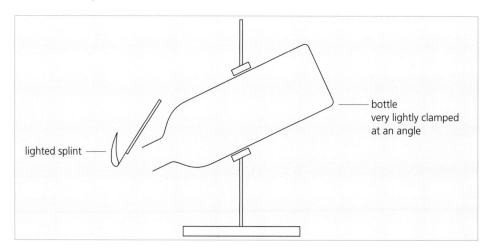

1. Very lightly clamp the filled bottle at an angle with the neck end pointing backwards.
2. Ensure the bottle will not hit any breakable objects – ideally fire it outside.
3. Light a splint attached to a metre rule.
4. Unscrew the bottle cap with the other hand – this can be done part-way in advance, securing the cap with a piece of masking tape.

5. **AT ARM's LENGTH** hold the tongs and splint to the open neck of the bottle. The gas mixture will explode, firing the bottle about 10 m.
6. Show that water is produced in the bottle – test with cobalt chloride paper – and that heat is produced. (However, able students will note that the inside of the bottle was wet initially.)

Advice:
1. The second bottle acts as a spare in case the gas mixture was not accurate. It can be difficult to judge when air has been displaced from the equipment leading to poor mixtures and a feeble result.
2. Prepare the bottles just before they are needed – do not store them overnight. This also produces a poor result as gases (especially hydrogen) diffuse through the plastic.

Answers

1. The energy of the reaction produces hot gases which caused the bottle to move away from the launch point very quickly. The bottle gets hot. There is a noise.

2. The energy is used to move the bottle.

3. Some water can be seen in the bottle (but this may be moisture from preparing the gases). We know from the equation that this produces water.

4. The water mainly goes into the atmosphere as water vapour.

5. (a) The energy would be used to power the engine, making the wheels turn and the car move.
   (b) The water would leave the car from the exhaust pipe and go into the atmosphere or be condensed.

6. Cars could not allow hydrogen and oxygen to combine in such an uncontrolled way, or it would be very dangerous! Also, we would need to be able to change speed and stop and start the reaction as we wanted.

## *Demonstration: Bubble trouble*

*Index 03.08*

Splitting water and making water in the same reaction

This demonstration uses electrolysis of dilute sulfuric acid to produce hydrogen and oxygen. The gases are bubbled into a soap solution and are then exploded.

The advantages of this demonstration are that the students see two reactions at the same time - this may save time. Disadvantages are that it is much harder to see the water produced in the hydrogen-oxygen reaction. Also, the reaction here is electrolysis of sulfuric acid – slight deception is needed in telling students this is the electrolysis of water.

### Apparatus and equipment
- 500 cm$^3$ conical flask with sidearm and rubber tubing attached
- Bung to fit flask with two holes to take the electrodes
- Power pack
- 2 leads with clips
- 2 pieces of 18 SWG solder wire electrodes, about 15 cm long, coiled
- Shallow dish
- About 50 cm$^3$ soap solution (**Minimal hazard**)
- 300 cm$^3$ 2 mol dm$^{-3}$ sulfuric acid (**Corrosive**)
- Plastic pipette to fit the rubber tubing
- Bunsen burner and heatproof mat
- Adhesive tape
- Eye protection
- Safety screen
- Splint.

### Safety
Wear eye protection.

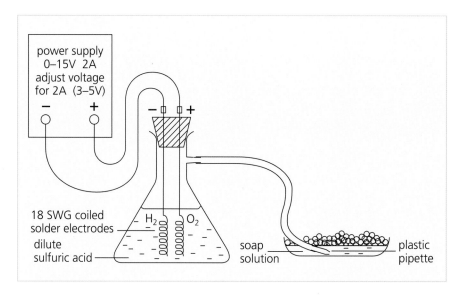

### What you do
1. Put soap solution in the dish to about two-thirds full.
2. Fit the solder wires through the rubber bung so the ends protrude by about 2 cm. Use adhesive tape to fix these in place if necessary.
3. Attach the plastic pipette to the rubber tubing and the other end of the tubing to the sidearm of the flask.
4. Put the sulfuric acid in the flask to about two-thirds full.
5. Fit the bung to the flask.
6. Connect the leads from the power pack to the electrodes.
7. Turn on the power supply to 3–5 V.
8. Allow the electrolysis to proceed for several minutes to ensure the rubber tubing is flushed of air. Bubbles should be seen at the electrodes in the flask.
9. Place the tip of the pipette into the soap solution.
10. Bubbles will form on the surface of the solution which can be lighted with a splint. To do this, first move the dish away from the flask and move bubbles away from the dish by sweeping off with a metre rule. Light the bubbles at arm's length.
11. It is also possible to make large bubbles which can be floated in the air and exploded. To do this, keep the end of the pipette in one place and grow large bubble on the end. Lift the end of the pipette in the air and allow the bubble to float off. **Light cautiously with the splint at arm's length.**

Answers

1. You know energy is released because you hear and see an explosion.

2. The energy is released when chemical bonds are made between hydrogen and oxygen atoms.

3. You cannot see the water – but you can predict it from the equation.

4. The water goes into the air as water vapour.

5. If the reaction happened in a car engine:-
   a) To power the car and make it move.
   b) It would come out of the exhaust or be condensed.

6. Cars could not generate the gases this way because it would be too dangerous to make an explosive mixture like this. Also the electrolyte would soon run out.

 ## *How much energy?*

*Index 03.09*

Answers

1. Four O-H bonds form in the reaction (the molecular models can be used to support this).

2. $4 \times 463 = 1852$ kJ $mol^{-1}$, so $-1852$ kJ $mol^{-1}$

3. Two H-H bonds and one O=O bond are broken: $864 + 498 = +1362$ kJ $mol^{-1}$

4. $1362 - 1852 = -490$, $-490 / 2 = -245$ kJ $mol^{-1}$

## *How much energy? (Alternative questions)*

Answers

1. $2H_2(g) + O_2(g) \rightarrow 2H_2O(l)$

2. Four O-H bonds form.

3. $4 \times 463 = 1852$ kJ $mol^{-1}$ , so $-1852$ kJ $mol^{-1}$

4. Two H-H bonds and one O=O bond are broken: $864 + 498 = +1362$ kJ $mol^{-1}$

5. $1362 - 1852 = -490$, $-490 / 2 = -245$ kJ $mol^{-1}$

## *Splitting water – making hydrogen and oxygen*
## *Demonstration*

*Index 03.10*

This is a useful way of illustrating the principle of producing hydrogen needed for fuel cells.

Two approaches are described – a demonstration using the Hoffmann apparatus and a class experiment.

Using the Hoffmann apparatus

Apparatus and equipment
- 1.5 dm$^3$ of 1 mol dm$^{-3}$ sodium sulfate solution (**Minimal hazard**)
- 60 cm$^3$ bromothymol blue indicator solution **OR** 60 cm$^3$ litmus indicator solution – both green when neutral (**Check suppliers sheets for hazard**)
- 10 cm$^3$ 0.1 mol dm$^{-3}$ sulfuric acid (**Minimal hazard**)
- 10 cm$^3$ 0.1 mol dm$^{-3}$ sodium hydroxide solution (**Irritant**)
- 2 x 2 dm$^3$ beakers
- Stirring rod
- Dropping pipette
- Hoffman apparatus
- Power pack
- Labels to stick to the Hoffmann apparatus
- White card, large enough to stand behind the Hoffmann apparatus
- Adhesive tape
- 2 test-tubes
- Splint
- Matches
- Bunsen burner and heatproof mat
- Eye protection
- Safety screen.

 Safety
Wear eye protection

What you do

Preparation
1. Put 1.5 dm$^3$ sodium sulfate solution into a 2 dm$^3$ beaker.
2. Add 60 cm$^3$ indicator solution. Stir.
3. Adjust the indicator colour to neutral (green for bromothymol blue) using the sulfuric acid or sodium hydroxide.
4. Close the stopcocks on the Hoffmann apparatus (this will prevent the gas escaping).
5. Fill the bulb with the coloured sodium sulfate solution.
6. Fill the arms of the Hoffmann apparatus in turn, by opening the stopcock part-way until the arm fills with the solution. Pour the solution in through the bulb in the centre. More solution may need to be added to the bulb. Close the stopcocks again.

7. Connect the leads from the power pack to the platinum electrodes at the base of the apparatus.
8. Label the negative terminal cathode and the positive terminal anode.
9. Place the white card at the back of the apparatus, so the coloured liquid is clearly visible to the audience.
10. Place a safety screen in front of the apparatus.

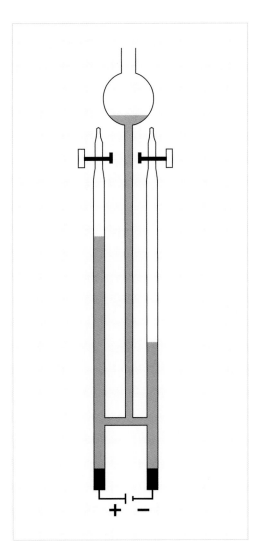

Presentation
1. Turn on the power supply to about 25 V DC – this produces a rapid reaction. Bubbles of gas will form in the arms of the apparatus.
2. Ask the students to observe the amounts of gas – there will be twice as much in the cathode arm as in the anode – and the colour of the liquid. If bromothymol blue is used, the anode solution will turn blue and the cathode solution yellow; for litmus the colours will be purple/blue and red respectively.
3. When the cathode arm is about two-thirds full of gas, switch off and disconnect the power supply.
4. Light the Bunsen burner and have a splint ready.
5. Hold a test-tube inverted over the tip of the anode arm. Ask the students to predict which gas is present.
6. Release the stopcock, allowing gas into the test-tube. Cover the end with a thumb. Light the splint.
7. Test the gas with the lighted splint, then release more gas and test with a glowing splint. This demonstrates the presence of oxygen.
8. Repeat the process with the cathode arm gas – a squeaky pop reveals that the gas was hydrogen.
9. Next, note the colours of the liquid.

Figure shows the Hoffmann apparatus at the end of the experiment – the hydrogen has been generated at the cathode clearly showing twice the volume of the oxygen gas generated at the anode.

Answers

1. (a) hydrogen (b) oxygen

2. The negative terminal (where hydrogen was produced).

3. Water → hydrogen + oxygen

4. The hydrogen could be used as a fuel to power the car. (You would need to collect the hydrogen and store it safely.)

Answers (for alternative questions)

1. Water has the formula $H_2O$. Therefore one molecule of water gives twice as many hydrogen atoms as oxygen atoms. As hydrogen and oxygen have the same formula ($X_2$) you always get twice as much hydrogen as oxygen.

2. The equations for the reactions at the electrodes are:-
   Cathode:      $2H_2O(l) \rightarrow O_2(g) + 4H^+ + 4e^-$
   Anode: $4H_2O(l) + 4 e^- \rightarrow 2H_2(g) + 4OH^-$

3. The overall equation for the reaction is found by adding these equations together:
$$2H_2O(l) \rightarrow O_2(g) + 4H^+ + 4e^-$$
$$4H_2O(l) + 4e^- \rightarrow 2H_2 (g) + 4OH^-$$
   _____
$$6H_2O(l) + 4e^- \rightarrow O_2(g) + 4H^+ + 4e^- + 2H_2(g) + 4OH^-$$

   Deleting the electrons on both sides and adding together the ions to make water molecules gives:
   $6H_2O(l) \rightarrow O_2(g) + 2H_2(g) + 4H_2O(l)$

   Deleting the excess water gives:
   $2H_2O(l) \rightarrow O_2(g) + 2H_2(g)$

   The colour changes of the indicators are due to the presence of hydroxide ions at the anode (alkaline) and the hydrogen ions at the cathode (acidic).
   The cathode reaction is oxidation (loss of electrons) and the anode reaction is reduction (gain of electrons).

4. This is the reverse of the equation for the 'rocket' demonstration.
   $2H_2(g) + O_2(g) \rightarrow 2H_2O(l)$

5. Electrolysing water can provide hydrogen for use in fuel cells. Energy is needed to break the -O-H bonds. In practice, this energy is likely to be from the Sun. Using solar power makes producing hydrogen in this way truly renewable, since the water formed in the fuel cell reaction can be recycled to produce the hydrogen again with no extra energy cost. Note that the water in a fuel cell must be pure – additional atoms, ions or molecules could poison the membrane.

## *Alternative demonstration method: using a dish on an overhead projector (OHP)*

This method is useful if no Hoffmann apparatus is available, but it is difficult to collect the gases. Showing the demonstration on the OHP is attractive.

### Apparatus and equipment
- Overhead projector
- 50 cm$^3$ 1 mol dm$^{-3}$ sodium sulfate solution (**Minimal hazard**)
- 2 cm$^3$ bromothymol blue indicator **OR** litmus indicator solution (**Check suppliers sheets for hazard**)
- 1 cm$^3$ 0.1 mol dm$^{-3}$ sulfuric acid (**Minimal hazard**)
- 1 cm$^3$ 0.1 mol dm$^{-3}$ sodium hydroxide solution (**Irritant**)
- 150 cm$^3$ beaker
- Stirring rod
- Dropping pipette
- 10 cm diameter crystallising dish, about 3 cm deep
- 2 platinum electrodes – 20 gauge wire, each about 8 cm long
- Adhesive tape
- Power pack
- Eye protection.

### Safety
Wear eye protection

Preparation
1. Pour 50 cm³ sodium sulfate solution into a 150 cm³ beaker. This is to produce a strong electrolyte.
2. Add 2 cm³ indicator solution. Stir. Use the sulfuric acid or sodium hydroxide solutions to adjust the solution to the neutral colour (green for bromothymol blue).
3. Bend the platinum wires over the dish, opposite to each other. Make sure one end of the wire reaches the base of the dish, but that the wires do not touch. Tape the wires in place on the outside. Make sure the wires run flat against the surface on the outside of the dish.

Presentation
1. Place the dish on the OHP.
2. Connect the platinum wires to the leads from the power supply.
3. Fill the dish about half full with sodium sulfate solution.
4. Turn on the power supply and adjust the voltage to about 20 V.
5. Bubbles form at the electrodes and the colour of the liquid will change – to blue around the anode and yellow at the cathode (if bromothymol blue is used) and purple/blue and red respectively if litmus is used. Stirring the liquid will restore the initial colour briefly.
6. Disconnect the power supply. Carefully swirl the dish, mixing the colours – the solution will return to the original state.

(Both procedures are based on those described in *Chemical Demonstrations, A Handbook for Teachers of Chemistry, vol 4*, by Bassam Shakashiri, p 156–162, The University of Wisconsin Press, 1992.)

## *Splitting water – making hydrogen and oxygen*
## *Class experiment*

*Index 03.11*

This experiment is satisfying for a class, but is quite fiddly, requiring good manual dexterity. The platinum electrodes can be replaced by carbon rods, however, as oxygen is formed some disintegration may occur due to a reaction with the carbon electrodes to form carbon dioxide, which causes soiling to the solution.

Note an alternative experiment – teachers might like to consider using a similar experiment described in a previous RSC Teacher fellow publication that also goes on to illustrate how to generate electricity. Refer to D. Warren, *Green Chemistry*, London: Royal Society of Chemistry, 2001 and see section Fuels for the future: How fuel cells work and Making a hydrogen fuel cell.

Apparatus and equipment (per group)
- 250 cm³ beaker or shallow dish, *eg* margarine tub (this is preferred if gases are to be collected)
- DC power supply or 6 V battery
- 3 connecting leads with crocodile clips
- 6 V bulb
- 2 platinum metal strips about 2 x 8 cm
- 150 cm³ tap water
- About 4 spatula measures (around 10 g) solid sodium sulfate (**Minimal hazard**)

- Bromothymol blue indicator, enough to make a strong colour (**Check suppliers sheets for hazard**)
- Few drops of 0.1 mol dm$^{-3}$ sodium hydroxide solution (**Irritant**) or 0.1 mol dm$^{-3}$ sulfuric acid (**Minimal hazard**)
- Eye protection.

To collect and test the gases
- 2 test-tubes
- Water
- Splint
- Bunsen burner and heatproof mat.

Safety
Wear eye protection

Instructions and notes
Full instructions are provided in the student section.

1. The sodium sulfate is present to create a strong electrolyte. Ask students to test the electrolytic behaviour of the water before and after adding the salt. Explain that the ions generated from the dissolved salt conduct electricity, so enable the water molecules to be electrolysed. This discussion could be taken further to differences in electrode potentials between the water molecules, sodium ions and sulfate ions.
2. The indicator colour should change to blue at the anode (alkaline) and yellow at the cathode (acidic). This should help students work out the reactions.
3. Mixing the solution in the beaker after switching off should produce the green colour again.

Observations
- Students should see that the colour of the indicator changes to blue around the anode and yellow around the cathode. Bubbles of gas will be produced at both electrodes – collecting and testing these will reveal oxygen at the anode and hydrogen at the cathode.
- The relative amounts of gas may be determined with care – there should be twice as much gas at the cathode than at the anode.
- The hydrogen gas will burn with a squeaky 'pop' and the oxygen gas will relight a glowing splint.

Answers

1. (a) hydrogen (b) oxygen

2. The negative terminal (where hydrogen was produced).

3. Water → hydrogen + oxygen

4. The hydrogen could be used as a fuel to power the car. (You would need to collect the hydrogen and store it safely.)

Answers (for alternative questions)

1. Water has the formula $H_2O$. Therefore one molecule of water gives twice as many hydrogen atoms as oxygen atoms. As hydrogen and oxygen have the same formula ($X_2$) you always get twice as much hydrogen as oxygen.

2. The equations for the reactions at the electrodes are:-
   Cathode:      $2H_2O(l) \rightarrow O_2(g) + 4H^+ + 4e^-$
   Anode: $4H_2O(l) + 4e^- \rightarrow 2H_2(g) + 4OH^-$

3. The overall equation for the reaction is found by adding these equations together:
   $2H_2O(l) \rightarrow O_2(g) + 4H^+ + 4e^-$
   $4H_2O(l) + 4e^- \rightarrow 2H_2(g) + 4OH^-$

   _____

   $6H_2O(l) + 4e^- \rightarrow O_2(g) + 4H^+ + 4e^- + 2H_2(g) + 4OH^-$

   Deleting the electrons on both sides and adding together the ions to make water molecules gives:
   $6H_2O(l) \rightarrow O_2(g) + 2H_2(g) + 4H_2O(l)$

   Deleting the excess water gives:
   $2H_2O(l) \rightarrow O_2(g) + 2H_2(g)$

   The colour changes of the indicators are due to the presence of hydroxide ions at the anode (alkaline) and the hydrogen ions at the cathode (acidic).
   The cathode reaction is oxidation (loss of electrons) and the anode reaction is reduction (gain of electrons).

4. This is the reverse of the equation for the 'rocket' demonstration.
   $2H_2(g) + O_2(g) \rightarrow 2H_2O(l)$

5. Electrolysing water can provide hydrogen for use in fuel cells. Energy is needed to break the -O-H bonds. In practice, this energy is likely to be from the Sun. Using solar power makes producing hydrogen in this way truly renewable, since the water formed in the fuel cell reaction can be recycled to produce the hydrogen again with no extra energy cost. Note that the water in a fuel cell must be pure – additional atoms, ions or molecules could poison the membrane.

## *Hydrogen fuel cells*

This section aims to show students what hydrogen fuel cells are like. This can be done using the text and pictures provided. If possible, it is a good idea to purchase a fuel cell and use this to illustrate the points.

Refer also to D. Warren, *Green Chemistry*, London: Royal Society of Chemistry, 2001 and see section Fuels for the future: How fuel cells work and Making a hydrogen fuel cell for additional activities.

# Future fuels

Learning objectives
- What a hydrogen fuel cell is and how it works
- Finding out the equations for the electrolysis of water
- Seeing how the electrolysis of water is applied to generating electrical power
- Considering how fuel cell technology may be applied to cars.

Time required
About 40 minutes for the written task alone.

A few more minutes will be needed for a basic demonstration of the fuel cell car.

More time can be devoted to the topic if the fuel cell kits are purchased – the texts with the kits have full instructions and more experiments.

If a fuel cell car is available, this can be charged up while students are working on the written task. Alternatively, the charging can be done before the lesson, and the time devoted to working out how the car works. It is a very intriguing piece of equipment.

Note that students may think that storing hydrogen in a car is much more dangerous than petrol. This is not really true – petrol is a highly flammable, toxic liquid made from a mixture of chemicals. BMW's tests indicate that their hydrogen car has identical safety potential to their existing models.

Answers

1. The energy is released in the form of an electric current.

2. The energy is used to power the car.

3. A full-size system would need a tank of hydrogen, a larger membrane surface and a larger, more powerful motor. Oxygen would be taken in from the air.

4. The only exhaust gas is water vapour. Although this is a greenhouse gas, this is the only one produced, so there is much less pollution than with a conventional engine. Yes, this will reduce the greenhouse effect.

5. **Good points** - the reduction in emissions, less maintenance (fewer moving parts), no shortage of fuel, probably lighter engines so fewer materials used in car manufacture.
   **Bad points** - needing to carry hydrogen, fuel cell membrane is very sensitive to poisons, so may need replacing and be expensive, hydrogen is not available as a fuel.

6. How to transport and store hydrogen gas safely and in a liquid format. How to get hydrogen gas into cars. How to store hydrogen in a car.

Resources

A variety of fuel cells are available for school use at a range of prices. Sources are suggested and teachers are advised to explore these.

In preparing this resource, the two kits from Heliocentris were purchased. Heliocentris kits are available from:
Economatics (Education) Ltd, Epic House, Darnall Road, Attercliffe, Sheffield S9 5AA

Telephone: + 44 (0) 114 281 3344
email: *education@economatics.co.uk*
website: *www.economatics.co.uk/education* (accessed November 2003)

Economatics sell two kits. The Hydrogen fuel cell car (£85.95 + VAT at the time of writing) includes a chassis and fuel cell which can be charged up using a solar cell. The solar cell is not included, although the kit comes complete with a small bottle of distilled water and an instruction book.

The Solar-hydrogen science kit is more expensive (£155 + VAT) but includes a larger fuel cell, a solar cell, a stopwatch and a load unit with a motor and an LED. Four different experiment books are included and the kit comes in a sturdy plastic container with lid, so the whole kit is self-contained.

Both kits are very high quality products which work well.

Kits are also available from Electro-Chem-Technic. The company sells a range of student-sized fuel cells which use liquid fuels such as a solution of sodium borohydride ($NaBH_4$). Their 'mini fuel cell' cost £24 at the time of writing, £20 each if four or more are ordered.
Electro-Chem-Technic, 81, Old Road, Headington, Oxford OX3 7LA
Fax: +44 (0) 1865 434799  (no telephone number available)
email: *info@ectechnic.co.uk*
website: *www.ectechnic.co.uk* (accessed November 2003)

*www.fuelcells.org* (accessed November 2003)
This is a very useful website which has extensive information about the topic, including links elsewhere.

## *Fuel cell cars: the GM AUTOnomy*

*Index 03.13*

**Learning objectives**
- That fuel cell technology is being applied to cars
- Cars may change how they work and function
- How to extract information from a press release.

**Time required**
About 40 minutes

In this activity, students look at a completely new concept in cars proposed by GM. This is the only car so far to have fully grasped the potential offered by hydrogen fuel cells. The AUTOnomy uses a chassis with interchangeable car bodies. The chassis can be of different lengths to take longer or shorter bodies and thus different numbers of passengers. The bodies can be of different types according to the circumstances in which the car is used.

The car is powered by hydrogen fuel cells. The driving point can be changed, as it will be computer-directed, not linked to a mechanical device. This is called the 'X-drive'.

The name AUTOnomy is exactly like the word 'autonomy' which is a way of saying independence. In other words, by driving AUTOnomy, a person is showing that they are free from the constraints imposed by conventional cars, and have a vehicle they can adapt to their needs when they want.

Answers

1. Second paragraph: 'Fuel cells' and 'X-by-wire' technologies. The 'X' means 'drive' or 'fly', so for a car this would be 'drive-by-wire'. The 'X-by-wire' technology is also being developed for use in helicopters.
   AUTOnomy is special because this is the first time the two technologies have been combined in a car.2. An AUTOnomy driver could have more than one car- a sports model, a family car and an estate car. The bodies could be stored in a garage and changed when required by the owner.

3. The sections appear in paragraphs 4 and 5.

4. Possible arguments include:
   **Agree**: Fuel cells are the way forward for powering cars of the future; rich countries of the world have no right to keep technologies just for themselves.
   **Disagree**: We can still use petrol for about the next 60 years; fuel cells are not the only solution (other fuels such as methanol are being developed as car fuels); if the all the world's population got cars this would drain resources and destroy habitats to build roads.

5. This is a good discussion point. We expect a certain standard of living in a developed nation, but what implications would arise if everyone had the same? If all families in the world had petrol-driven cars, this would use up the remaining supplies of crude oil very quickly. If increasing availability of cars is a global aim, then these should be zero emission. The problems facing the world's poor are much more basic than having a car or not, so it would be sensible to solve these first. The possibility here is to speculate on what would happen if the number of petrol-driven cars increased dramatically.

## *Further information, references and source material*

Other companies with fuel cell developments in the UK include:

Johnson Matthey Fuel Cells
200 Brook Drive, Green Park, Reading, Berks RG2 6UB
Tel: +44 (0) 118 949 7095
Fax: + 44 (0) 118 949 7295
Email: *fuelcells@matthey.com*
Website: *www.matthey.com* (accessed November 2003)

Further information about fuel cells, the AUTOnomy and the latest technological developments can be found on the Newsweek website at www.msnbc.com. (accessed August 2003). To find the articles, click on Tech&Science, then select Find on the right hand side. This permits a search with 'hydrogen fuel cells' as the parameters.

Further information about how GM are developing fuel cell cars can be found on the GM website at *www.gm.com* (accessed November 2003).

The Chemical Industries Association website at *www.cia.org.uk* (accessed November 2003) also has information about hydrogen fuel cells. search under hydrogen fuel cells to access a range of articles.

The websites cited in the students' material are:

*www.bmweducation.co.uk* (accessed November 2003) for hydrogen cars.

*www.ford.co.uk* (accessed August 2003) for petrol engine cars.

*www.toyota.com/html/shop/vehicles/prius/* (accessed November 2003) for the Prius – Toyota's hybrid car available in the UK.

*www.thecarbontrust.co.uk* (accessed November 2003) gives press releases and information about low carbon technology.

*www.actionenergy.org.uk* (accessed November 2003) focuses on reducing energy usage.

*www.dft.gov.uk* (accessed November 2003) is the website for the Department for Transport. The document 'Powering Future Vehicles Strategy' shows what the current government's plans are. To find this, click on 'Roads, Vehicles and Road Safety' then choose 'Vehicles' and then 'Cleaner vehicles'.

*www.vcacarfueldata.org.uk* (accessed November 2003) is a website dedicated to providing information about cars, emissions and other legal matters. It provides an easy-to-use search option to find information about any specific car model.

*www.est-powershift.org.uk* (accessed November 2003) is the website of the Energy Saving Trust. This has a lot of useful information about alternative fuel sources for cars, including the network of LPG suppliers in the UK. There are also links to other useful sites.

*www2.exxonmobil.com/corporate/* (accessed November 2003) is the Esso website.

Articles on fuel cells

C. Curran, *Fuel cells – alternative energy? Chemistry & Industry*, 3rd December 2001, 23 , 767.

R. Massey, *Stepping on the gas* Daily Mail 19th April 2002, 23.

M. Laughton, *Fuel cells* Power Engineering Journal February 2002 p 37–47
This is a very useful technical article which shows a wide range of different types of fuel cells.

Additional material on climate change and the greenhouse effect and fuel cells can be found in:

D Warren, *Climate change*, London: Royal Society of Chemistry, 2001.

D. Warren, *Green chemistry*, London: Royal Society of Chemistry, 2001.

**RS•C**    Future fuels

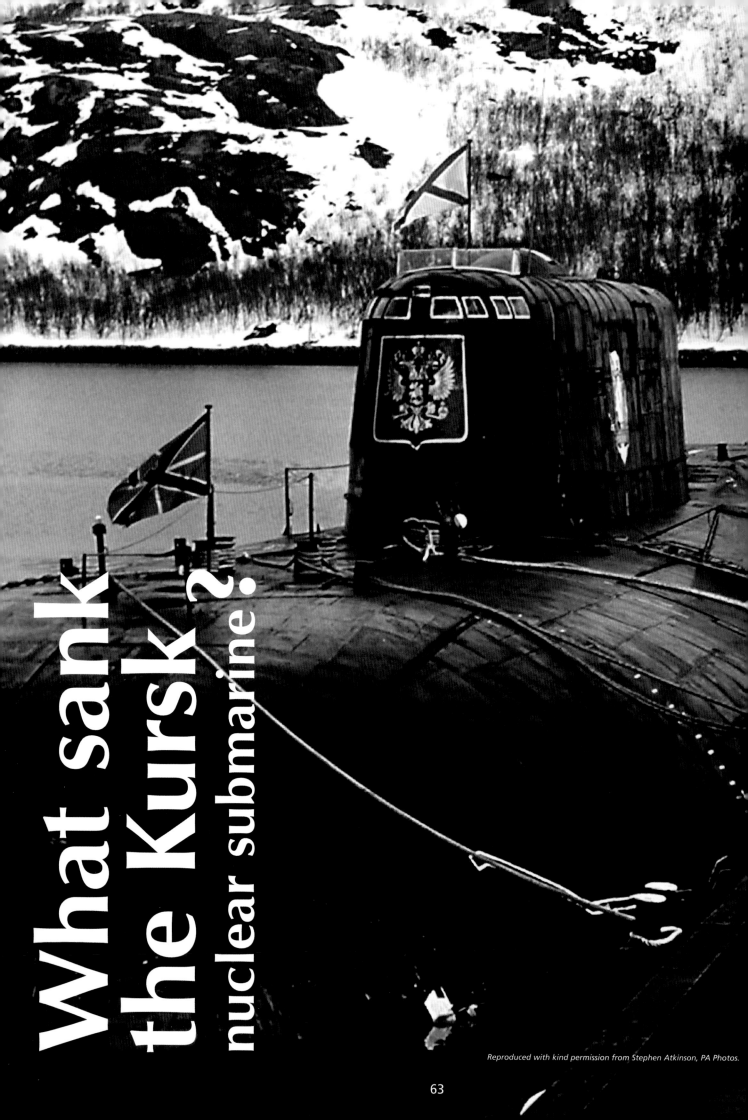

# What sank the Kursk?
## nuclear submarine?

*Reproduced with kind permission from Stephen Atkinson, PA Photos.*

# What sank the Kursk nuclear submarine?

Summary

The Russian submarine Kursk sank in August 2000 with the loss of 118 lives. This was a major news story, causing Russia great international embarrassment. This material explores the cause of the sinking.

| Resource name | Index | Type | Age range | Topic | Media |
|---|---|---|---|---|---|
| *Analysing the accident* | *04.01* | Paper based exercise – data analysis | 14–16 | Data analysis | |
| *Did you know? About seismology* | *04.02* | Information | 14–16 | Seismic analysis | |
| *What type of seismic disturbance was recorded?* | *04.03* | Paper based exercise – data analysis | 14–16 | Data analysis | |
| *Person profile – David Bowers* | *04.04* | Careers information | 11–16 | Careers | |
| *News timeline* | *04.05* | Information | 11–16 | News reports on the Kursk | |
| *Did you know? About hydrogen peroxide* | *04.06* | Literacy / DART | 14–16 | Hydrogen peroxide | |
| *Which catalyst? Decomposing hydrogen peroxide* | *04.07* | Demonstrations and questions | 14–16 | Decomposition of hydrogen peroxide, effect of catalysts on rate | |
| *Which catalyst? Decomposing hydrogen peroxide* | *04.08* | Class investigation experiment and questions | 14–16 | Decomposition of hydrogen peroxide, effect of catalysts on rate | |
| *Peroxide power in torpedoes* | *04.09* | Literacy / DART and questions | 14–16 | Hydrogen peroxide | |
| *Report: What sank the Kursk?* | *04.10* | Report writing exercise using information from different sources | 14–16 | Presenting information | |

Key: Interactive student activity    Photocopiable and printable worksheet    Projectable picture resource

# What sank the Kursk nuclear submarine?

### Issue

- A simple chemical reaction contributed to the sinking. Is chemistry responsible only for 'bad' events?

### Chemical topics

- Catalysis: the catalytic decomposition of hydrogen peroxide.

### Scientific enquiry issues

- Using scientific evidence to explain an event
- Investigating a chemical reaction
- Communicating findings
- Debating the ethical issues of reporting chemical science.

### Notes on using the resource

In this series of activities students find out how a simple chemical reaction contributed to the sinking of the Kursk submarine. Essentially, the Kursk submarine sank because two explosions occurred in the torpedo storage area in the bows. The first is likely to have been caused by leakage of highly concentrated hydrogen peroxide, while the second was probably one or more torpedoes exploding and was large enough to blow the front off the vessel. The submarine was flooded with water within hours. Although some crew members escaped to a rear compartment, most were killed by either the explosions or the fire which followed. After a few hours all the 118 crew were dead. The reasons for the explosions were chemical – the torpedoes used high test peroxide (HTP) in the firing mechanism. HTP was the propellant as liquid fuel was cheaper than solid fuel. This system was trialled in the UK Royal Navy torpedoes until the HMS Sidon disaster of 1955, in which 13 sailors lost their lives, in an accident similar to that which sank Kursk.

The unit begins with the seismic analysis done by leading UK seismologists of the traces produced by the explosions. Students can see that the Kursk could not have sunk due to an accident with a US or British submarine, but that underwater explosions were likely to be responsible. In the next section, they can carry out experiments or watch demonstrations using hydrogen peroxide and find out about the HTP torpedoes used on board Kursk. This set of activities is interlinked and it is recommended that students work through them in sequence. **Analysing the accident, What type of disturbance was recorded?** and **News timeline** could all be done as homework prior to the experimental activities, which explore the catalytic decomposition of hydrogen peroxide. The chemical reaction can be investigated either as a class experiment with varying degrees of open-endedness, or by using two demonstration reactions to illustrate key principles. Students are encouraged to write up their findings from activities in the style of a media reporter – again, variation is possible here.

Parts of the resource can also be used as stand alone material when dealing with rates of reaction or catalysis.

## Data analysis

Learning objectives
- To analyse seismograms to get information about the sinking of the Kursk
- To realise that the Kursk sank due to two underwater explosions
- To find out how seismologists work.

## Analysing the accident

*Index 04.01*

Time required
About 40 minutes

Apparatus and equipment
- Access to an atlas or to an electronic map

In **Analysing the accident** and **What type of disturbance was recorded?**, students use basic principles of seismology to find out about what must have caused the Kursk to sink. The seismological information is presented in three stages – first, students work out when and where the sinking occurred, next they find out what type of disturbance caused the sinking by analysing the size and shape of the traces and thirdly they work out the size of the explosions. This is a good exercise in interpretation and evaluation of data. Note however, that seismology alone cannot rule out collision with another submarine.

These activities are simplified versions of the actual steps used by seismologists to work out what happened to the Kursk. The story created a great deal of interest partly because of the human tragedy involved, but also because the Russians' explanations for the events did not tie with the data. This made the work of the seismologists important in establishing the true cause of the sinking.

Answers

1. The second disturbance is the bigger disturbance.

2. The disturbances occurred in the Barents Sea, north of Norway.

3. a) The Russians said that the Kursk had collided with another submarine.

   b) The BBC/other reporters suggested that there had been an accident involving a nuclear submarine.

## *Did you know? About seismology*

*Index 04.02*

## *What type of seismic disturbance was recorded?*

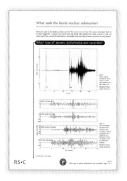  

*Index 04.03*

**Time required**
About 40 minutes

Traces from underwater explosions have a characteristic wave pattern called a 'bubble pulse'. This is consistent with a volume of gas expanding and contracting while moving up through the sea until breaking the surface. The Kursk traces, when produced in a different format to that used in the first exercise, are typical of an underwater explosion.

The other key point is that both the disturbances were underwater explosions, as both traces were of similar shape. This meant that other theories about Kursk having collided with a foreign submarine, or with the bottom of the sea, or with an old World War II mine was unlikely.

Answers

1. The time interval is 135 seconds – *ie* two minutes and 15 seconds.

2. The first Kursk signal is tiny compared to the second one.

3. The shapes of the two signals are the same. This suggests they were both caused by the same kind of source.

4. Figures 2 and 3 suggest that the signals were caused by underwater explosions.

5. The first disturbance might have been an explosion inside the submarine.

6. No, because the seismic data suggests that an underwater explosion was the cause, not a collision.

7. This saved them from being embarrassed about causing the accident themselves. Also, they probably could not believe that the Kursk could have sunk of its own accord.

8. The Richter magnitudes are 1.5 and 3.5 from Table 1 in worksheet 04.03.

9. It could absorb/modify them.

10. They are likely to have been larger.

## *Person profile – David Bowers*

*Index 04.04*

## *News timeline*

*Index 04.05 (5 pages)*

## *Peroxide power*

Learning objectives
- That hydrogen peroxide decomposes to form water and oxygen gas
- That the decomposition rate can be increased by catalysis
- That the decomposition of hydrogen peroxide is linked to the Kursk sinking.

# RS•C

## What sank the Kursk nuclear submarine?

Two alternatives are given – key features of the decomposition reaction can be shown in two demonstrations, or as a class experiment. The purpose of the experiments is to show how hydrogen peroxide reacts. In the third section, students are shown that a highly concentrated form of the chemical was used in torpedoes on the Kursk.

### *Did you know? About hydrogen peroxide?*

*Index 04.06*

### *Which catalyst? Decomposing hydrogen peroxide*

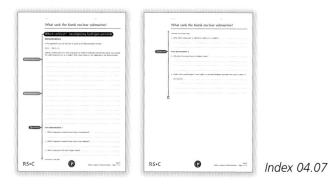

*Index 04.07*

Time required
About 20 minutes, depending on the length of discussion

### Demonstration 1:

Time required
About 10 minutes

In this demonstration the decomposition reaction:

$2H_2O_2(l) \rightarrow 2H_2O(l) + O_2(g)$

is done in the presence of washing-up liquid, which creates a foam. Varying the catalyst creates differences in the rate at which the foam is produced. The reaction instructions have been adapted from T. Lister, *Classic Chemistry Demonstrations*, London: Royal Society of Chemistry, 1995 (p 145 –146).

Apparatus and equipment
• One 250 cm$^3$ measuring cylinder for each catalyst tested
• Tray with raised sides to catch any spillage, one for each reaction

- Stopclocks – one for each reaction
- 50 cm$^3$ 100 volume hydrogen peroxide per reaction – **Corrosive**
- About 0.5 g of each solid compound to be tested (see list)
- A few drops of washing up liquid
- 2 dropping pipettes
- Bunsen burner, mat and splints
- Eye protection – goggles.

Compounds to test
- powdered manganese(IV) oxide – **Harmful**
- lead(IV) oxide – **Toxic**
- iron(III) oxide
- aqueous $Fe^{2+}$ solution
- aqueous solutions of ions of other transition metals.

Safety
Wear goggles.

Before the demonstration
1. Test the strength of the hydrogen peroxide carefully. If the reaction is too vigorous, dilute the concentration by 50% by adding 50 cm$^3$ water to each measuring cylinder.
2. Set each measuring cylinder in a tray.
3. Pour the hydrogen peroxide into each measuring cylinder.
4. Add a few drops of washing up liquid to each measuring cylinder.
5. The demonstration should be carried out in a well ventilated laboratory.

Carrying out the demonstration
1. Invite a student to start and time each reaction – two students will be needed per reaction.
2. Give one student the catalyst and the other the stopclock.
3. The student timing must start when the catalyst meets the hydrogen peroxide and stop when the foam reaches the top of the cylinder.
4. Line up all the students with the reactions.
5. Count down to the addition of the catalyst – try to ensure everyone adds their compounds simultaneously. Note that students adding aqueous solutions will need dropping pipettes.
6. Test the gas in the foam with a lighted splint – this should burn more brightly. Alternatively, test with a glowing splint which should relight. This will confirm the presence of oxygen.

Results

The lead(IV) oxide will probably be fastest followed by the manganese(IV) oxide.

Discussion about the differences could focus on the different surface areas – adding a solution promotes an even faster reaction and the powders may be made from particles of different sizes. Show that the powders are not changed in the reaction – they can be seen at the bottom of the cylinders after the reaction.

Discuss also that the powders act as surface catalysts providing a surface on which hydrogen peroxide molecules can stick.

## Demonstration 2:

Time required
About 10 minutes

# What sank the Kursk nuclear submarine?

In this reaction, the heat generated in the decomposition reaction can be appreciated clearly. 100 volume hydrogen peroxide is decomposed in the presence of manganese(IV) oxide to produce a cloud of water vapour from a narrow-necked flask covered in foil. The effect can be likened to a genie coming out of a bottle. The instructions have been adapted from V. Kind and R. Hadi-Talab *School Science Review* (in press 2003).

### Apparatus and equipment
- 500 cm$^3$ narrow-necked conical or flat-bottomed pyrex glass boiling flask with a tight-fitting stopper
- Aluminium foil
- About 30 cm fine invisible nylon thread
- Empty teabag, cut along one edge
- Enough manganese(IV) oxide to fill the teabag
- 40 cm$^3$ of 100 volume hydrogen peroxide – **Corrosive**
- Eye protection – goggles.

### Safety
Wear eye protection – goggles. Use safety screens.

### Before the demonstration
1. Fill the teabag with the manganese(IV) oxide and tie it closed using the nylon thread. Shake the teabag to remove any excess/small particles of $MnO_2$ falling through the pores before placing it in the flask. Leave a long piece of nylon thread at one end.
2. Cover the flask completely with foil.
3. Put the hydrogen peroxide in the flask.
4. Suspend the teabag in the flask so it hangs above the hydrogen peroxide.
5. Secure the long end of the thread in place with the stopper.

### Carrying out the demonstration
1. Tell the students that you have a genie in the flask desperate to escape.
2. Invite a student to release the genie from the bottle. S/he has to make a wish and stroke the bottle three times, keeping the flask on the bench.
3. After the third stroke, discreetly pull out the stopper and stand back.
4. After about two seconds, a jet of steam and oxygen will be generated from the neck of the flask, forming a cloud in the air. This is the genie. The flask will be quite hot after the reaction.

### Points to consider
The decomposition reaction can create a high temperature sufficient to vapourise the water to steam. This is important – imagine the power that could be created from the simple reaction! Controlling the release of the gases would be like controlling a powerful explosion. This was realised by rocket and torpedo scientists working in the 1930s (see resources for further information).

### Answers

### From demonstration 1

1. This is likely to be lead(IV) oxide or manganese(IV) oxide.

2. This is likely to be the iron(III) oxide.

3. The compound producing the foam most quickly.

4. Encourage students to think of other transition and precious metals. Silver is a good catalyst.

From demonstration 2

5. The decomposition is occurring at sufficiently high temperature to vapourise the water. This would happen when highly concentrated hydrogen peroxide is used.

6. The gases would be produced at very fast rates and at high temperatures, rather like an explosion.

## *Which catalyst? Decomposing hydrogen peroxide*

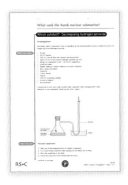

*Index 04.08*

Investigation

**Time required**
This depends on how open ended the investigation will be. As described, about 60 minutes would be needed.

In the method described, the test samples are placed in small tubes which are held in place within a larger flask until timing of the reaction is ready to start. This helps to create a more accurate starting time, but requires more labour in preparation. As an alternative, the catalyst compounds could be placed at the bottom of each flask to start with and the hydrogen peroxide simply poured in. In this case, timing should start when half of the hydrogen peroxide has been added.

The open-endedness of the task can be varied depending on the time available. For example, students could be given only an apparatus list and invited to work out the experimental method using this, or they could be given the title of the investigation only. Students could be invited to research the reaction before starting, having the opportunity to plan their own work. The demonstrations above could be used as the basis for this.

There are other ways of collecting the gas, for example using gas syringes, or by adapting the method used in demonstration 1 in which foam is measured. Students could be encouraged to devise their own system, depending on the facilities available.

Discussion points will be the amounts of materials to use. In practice, much smaller amounts of catalyst are required than students think – only about 0.3-0.5 g are needed.

Manganese(IV) oxide is likely to be the most effective catalyst.

**Apparatus and equipment (per group of students)**
- Trough
- 50 cm$^3$ burette
- 250 cm$^3$ conical flask with sidearm and bung fitted

# What sank the Kursk nuclear submarine?

- About 75 cm³ of 20 volume hydrogen peroxide per test – **Irritant**
- Range of compounds to test – see list for suggestions
- Small test-tubes / sample tubes
- Rubber tubing to connect sidearm to mouth of burette
- Boss, clamp and stand
- Stopclock
- Cotton thread
- Funnel
- 100 cm³ measuring cylinder
- Access to balance
- Zinc oxide
- Iron(III) oxide
- Copper(II) oxide – **Harmful**
- Manganese(IV) oxide – **Harmful**
- Powdered metals *eg* zinc, iron, copper
- Eye protection.

Safety
Wear eye protection.

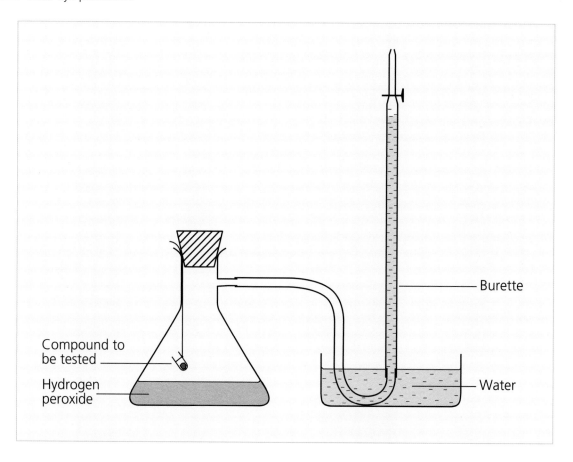

Compound to be tested

Hydrogen peroxide

Burette

Water

Answers

1. Manganese(IV) oxide is likely to be the best catalyst of those listed.

2. Other results will depend on those used, but iron(III) oxide is likely to be the least effective.

3. The decomposition reaction would occur at a very fast rate, like an explosion.

# RS•C

## Peroxide power in torpedoes

*Index 04.09*

### Learning objectives
- Extracting information from unfamiliar text.

### Time required
About 30 minutes

This activity can be used to follow on from either experiment or demonstrations. It shows that the decomposition of hydrogen peroxide has been used in torpedo technology. Students could carry out further research using the information provided at the end of the unit, as the extracts given here are necessarily short.

### Answers

1. An explosion occurred when a torpedo launch was attempted. This involved HTP leaking onto metal and decomposing so fast that the gas build-up caused an explosion.

2. They were considered too dangerous after the Sidon accident.

3. The Red Star stated that the Kursk was carrying HTP-based torpedoes.

4. The Kursk had been refitted in a 'modernisation' programme designed to save money.

5. The story was probably removed because it was too near the truth and at the time the Russian Navy did not want to admit to this.

6. The torpedo designer thinks that the Kursk and Sidon accidents were caused by the same thing – leakage of HTP followed by an explosion due to the build-up of gas.

### *Report: What sank the Kursk?*

*Index 04.10*

Learning objectives
- To prepare a report suitable for the general public about a scientific issue
- To draw together information from several sources.

Time required
About 60 minutes as a minimum

At a basic level, students are invited to produce a report about the sinking of the Kursk in the style of a media reporter for a news agency, incorporating their analysis of the seismic traces, findings from experiments and research on torpedo technology. Alternatives include:
- carrying out an interview in which two students work together, one as the reporter and the other as the scientist. This could be recorded on video or tape;
- carrying out a role play as a 'Board of Inquiry' into the Kursk disaster, in which students take on roles, such as a Navy representative, expert scientist, Government official, submarine captain, torpedo designer; each makes their own case for what happened. See **Future fuels** (page 42) for advice on running a role play; and
- making a collective display of the findings, with teams of students working on different aspects such as the seismic analysis, chemical experiment and torpedo technology.

Students should be encouraged to carry out further research using the information list at the end of the unit. There is an extensive range of material freely available which would provide sound background for this type of activity.

## Further information, references and source material

These websites and URLs have information about the sinking of the Kursk. Typing Kursk in the search engines of any major news website will produce a list of articles relating to this event.

*www.bellona.no* (accessed November 2003) is a Norwegian environmental issues site run by the Bellona Foundation which specialises in events occurring in Russia, particularly the Northern Fleet. On entering the site, look for the UK flag, the Union Jack, in the top right hand corner of the home page if you prefer to read the material in English! The site has over 50 articles concerning the Kursk.

*www.bbc.co.uk/news* (accessed November 2003) is the BBC's main newsite carrying many articles about the sinking of the Kursk. *www.bbc.co.uk/science/horizon/2000/kursk.shtml* (accessed September 2003) is the location for the transcript of the Horizon programme called 'What sank the Kursk?' shown in August 2001.

*www.blacknest.gov.uk* (accessed November 2003) is the website for Blacknest, the section of the Atomic Weapons Establishment in the UK responsible for seismic analysis.

*http://members.aol.com/nicholashl/ukspace/htp/htp.htm* (accessed November 2003) is a webpage devoted to hydrogen peroxide rockets. Maintained by Nicholas Hill, a chemistry teacher, the page includes fascinating historic photographs of rocket production during the 1940s – 50s.

*www.observer.co.uk/focus/story/0,6903,532299,00.html* (accessed November 2003) is a special article exploring the reasons for the sinking. It can also be accessed through The Guardian newspaper site at *www.guardian.co.uk* (accessed September 2003) and typing 'unsinkable sub' in the search engine.

*http://kursk.strana.ru/english/* (accessed November 2003) is the official Russian website, in English, relating to the sinking of the Kursk. The site includes an interactive diagram of the submarine and a video taken of the destroyed vessel on the seabed, with an English commentary.

*www.janes.com/defence/naval_forces/* (accessed November 2003) is part of the website for Jane's, the famous surveyors of the world's navies. The site carries a number of well-researched articles on Kursk.

Other source material and references include:-

V. Kind  and R. Hadi-Talab, *Demonstrating Chemistry,* School Science Review (in press).

T. Lister, *Classic Chemical Demonstrations,*  London: Royal Society of Chemistry, 1995.

Hair

# RS•C    Hair

**Summary**
The look and style of hair is a major pre-occupation amongst young and many other people. We invest money in having hair cut, styled, coloured and treated. This material investigates the basics about hair.

| Resource name | Index | Type | Age range | Topic | Media |
|---|---|---|---|---|---|
| Shampoo survey | 05.01 | Paper based exercise – data collection | 11–14 | Data collection | |
| Shampoo survey results | 05.02 | Paper based exercise – data analysis | 11–14 | Data analysis | |
| Wash in style | 05.03 | Experiment | 11–14 | Testing shampoos | |
| Cosmetic ingredients database | 05.04 | Database | 11–16 | Database | |
| Key words | 05.05 | Glossary | 11–16 | Glossary | |
| Did you know? About hair | 05.06 | Literacy / DART | 11–14 | Structure of hair | |
| Hair and shampoo – the facts | 05.07 | Paper based literacy / DART | 14–16 post 16 | Hair structure and shampoos | |
| Making shampoo | 05.08 | Experiment | 11–16 | Making shampoo | |
| Results table for shampoo tests | 05.09 | Data collection | 11–16 | Data collection | |
| Shampoo in the shop | 05.10 | Group work analysis, discussion, project work and presentation | 11–16 | Consumer marketing | |
| Person profile – Rahila Begum | 05.11 | Careers information | 11–16 | Careers | |
| Bad hair day – some hair headaches | 05.12 | Paper based literacy / DART | 11–16 | Hair problems | |
| Did you know? The truth about dandruff | 05.13 | Information | 11–16 | Dandruff | |
| Fringe benefits – investigating shampoos | 05.14 | Written exercise – planning an experiment | 11–14 | Planning an experiment | |
| Did you know? Hair today, gone tomorrow | 05.15 | Information | 11–16 | The truth about baldness | |

Key:   Interactive student activity     Photocopiable and printable worksheet     Projectable picture resource

Issue
What do shampoos and conditioners actually do to the hair and do they really work as they say?

Chemical topics
- Making mixtures
- Action of surfactants on grease
- Measuring pH
- Protein in hair.

Scientific enquiry issues
- Collecting evidence when variables cannot readily be controlled
- Deciding the range of data to be collected and methods for testing
- Using scientific knowledge to decide on a strategy for an investigation
- Presenting conclusions to support predictions
- Presenting scientific data to the public.

Notes on using the resource

The activities are all free-standing, but most make sense when connected to one or more of the others. In particular, **Hair and shampoo – the facts**, a text-based activity, provides a lot of background for several others, so even if this activity is not used as a task, the sheets would be useful for students working on later activities *eg* **Shampoo in the shop**, **Bad hair day – some hair headaches** and **Fringe benefits – investigating shampoos**. The **Shampoo survey** makes most sense when connected to activity **Wash in style** which involves testing shampoos.

The activities have been designed with 11-16 year olds in mind, but the level of reading required in **Hair and shampoo – the facts** may mean that some help is required.

The **Cosmetic ingredients database** is also needed for **Wash in style**. It also provides a useful source of reference information for the text-based activity in **Hair and shampoo – the facts**.

The **Shampoo survey** provides an introduction to the topic by inviting students to find out which shampoos are used among people they know. These could be classmates or relatives, or a mixture of both. If each student asks about 8–12 others, there should be a good range of data within the group. This leads to the possibility of handling larger figures and compiling information across a wider range of people. The survey is deliberately very basic, but this does not preclude addition of other questions, such as the age range of respondents. The task can be developed further.

In **Wash in style**, students test shampoos in two ways; their pH values and the extent to which they disperse a pool of oil. The results are foolproof and the experiment could be extended to include the production of foam, a judgement of smell and so on. The **Cosmetic ingredients database** is needed to look up the components.

In **Hair and shampoo – the facts**, students can find out exactly what hair is and what happens when we wash it by working through the text-based activity. This contains a lot of information used elsewhere. The sheets could be used for reference only.

For **Making shampoo**, advice on reagent suppliers is provided. This is a very good activity and yields a very pleasant product. The notion of marketing a product is explored in **Shampoo in the shop**. Here, students can adopt roles in marketing a product and testing it. To make the most of this, teachers could arrange a 'launch' meeting at which groups present their 'products'. The shampoo made in **Making shampoo** may be a good choice for this.

In **Bad hair day – some hair headaches**, students help a hairdresser with her chemistry. Four fictional students' hair problems are described. Students have to explain the chemistry behind the problems. The information needed is gained from doing other activities, or from **Hair and shampoo – the facts**.

Finally, **Fringe benefits – investigating shampoo** offers an investigation into shampoos. This presents the opportunity to investigate manufacturers' claims about their products. Students can consider how manufacturers may justify their claims about shampoos and in particular how these could be tested.

## *Shampoo survey*

*Index 05.01*

Learning objectives
- To collect survey data
- To analyse survey data
- To calculate an average cost for shampoo
- To consider reasons for choosing shampoos.

Time required
- At least an hour if carried out using people outside school.
- If the information is collected within a class in a lesson then about 50 minutes is needed.

This is a very straightforward activity which could be set as a homework. The data collected could be used collectively to assess a larger group of people. Given the likely age of those undertaking the survey, a 'Name' column has been included on the **Shampoo survey results** sheet (Index 05.02), as it is unlikely friends and relatives would take offence at inclusion in this way. However, it would be worth discussing with students anonymity in surveys and the possible need to change 'Name' to a code number, with a list of numbers and names kept elsewhere. Students should also be reminded of the importance of collecting enough data.

Alternatively, teachers may like to consider allowing students to devise their own questions and results sheets allowing students to present their results in different forms *eg* bar charts/pie charts etc.

Answers

1–4. Answers will be dependent on survey results.

5. Students may give reasons such as price, 'its the family shampoo', doing sport, hairdresser recommended it.

6. The ingredients in the shampoos are supposed to be best suited to a particular hair type. To keep hair looking good it is therefore probably best to use a shampoo which matches hair type.

## Shampoo survey results

Index 05.02

## Wash in style

Index 05.03

---

**Learning objectives**
- To test the pH values of shampoos
- To test how shampoos compare in dispersing oil
- To compare the ingredients of different shampoos and relate these to the test results.

---

Time required
About 60 minutes
Allow about 5 minutes for each shampoo to be tested.

Apparatus and equipment (per group, assuming 6 tests)
- Universal Indicator solution (**Check supplier sheets for safety information**)
- About 200 cm$^3$ distilled water (this is enough for about 10 tests)
- 6 test-tubes
- Test-tube rack
- Dropping pipette – clean one for each shampoo
- 6 petri dishes or similar
- About 200 cm$^3$ cooking oil (the actual type of oil does not matter)
- Piece of dark (black) paper, about 10 cm x 10 cm to go under a dish
- Eye protection
- Access to the **Cosmetics ingredients database**
- Copy of **Results table** for each person
- Selection of shampoo smaples.

Results
The price 'per 100 ml' can be obtained from supermarket shelves or calculated from the price of the bottle of shampoo. Although strictly the volume unit should be 'cm$^3$', 'ml' is retained here to provide continuity with the information students are likely to find. The costs are based on January 2002 figures. There are many shampoos on the market, so it is impossible to test all of them. Try to get a range.

Answers

1. Shampoo dissolves dirt and grease.

2. Soap or detergent give washing properties, other chemicals can thicken the shampoo, control pH or act as a preservative.

3. Answers will depend on the products tested. The oil drop test results may vary. This is due to the other ingredients in the shampoo and therefore a logical conclusion may not be obtained. In basic laboratory tests all shampoos 'clean' approximately the same amount. Shampoos with thickeners tended to spread out less – students could discuss if adding a 'thickener' masks the action of the shampoo, adding only to the appearance and texture of the product.

4. Answers will depend on the products tested.

5. Answers will depend on the products tested, however, in general shampoos tend to have pH values in the range 6.0–7.5

6. Before shampoo people would just use water or ash from fire and water.

Sample results

| Name of shampoo | Cost per bottle/£ | Price per 100 ml (cm$^3$)/£ | pH value | Diameter of circle in oil test/cm |
| --- | --- | --- | --- | --- |
| Trevor Sorbie | 2.99 | 1.19 | 7.5 | 3.3 |
| Garnier Neutralia | 2.69 | 1.08 | 7.5 | 2.0 |
| Salon Selectives | 1.99 | 0.47 | 6.5 | 3.5 |
| Tesco Value | 0.34 | 0.34 | 6.0 | 4.4 |

Answers to the questions below the table will depend on the shampoo samples tested.

Use of the **Cosmetic ingredients database** will show the main ingredients added for the functions listed. Here are key names which students should find:

**Surfactant**    Sodium lauryl sulfate
Sodium laureth sulfate (milder)
Ammonium lauryl sulfate
Ammonium laureth sulfate

**Preservative**  Butylparaben
Benzyl alcohol
Benzophenone
Disodium EDTA
Citric acid

**Emulsifier**    Cocoamidopropyl betaine
Cocoamide DEA

**Salt**    Sodium chloride
Magnesium chloride

**Thickener**    Behenyl alcohol
Carbomer

**Perfume**    These are commonly fruit oils, *eg* citrus sinensis or more commonly 'Parfum', which means a mixture of different fragrances.

5-digit numbers in the list are for colouring pigments.

Further information is available in *A Consumer's Dictionary of Cosmetic Ingredients* by Ruth Winter and *Natural Health and Body Care* by Neal's Yard Remedies (see resources list).

## *Cosmetic ingredients database*

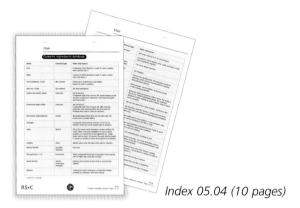

*Index 05.04 (10 pages)*

## *Key words*

*Index 05.05*

## *Did you know? About hair*

*Index 05.06*

## Hair and shampoo – the facts

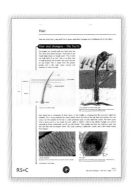 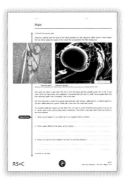 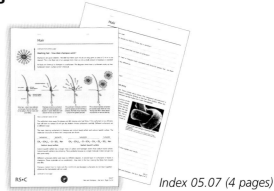

Index 05.07 (4 pages)

---

### Learning objectives
- To understand what hair is made from
- To understand why hair gets dirty
- To understand how shampoos and conditioners work
- To develop understanding about chemical bonding.

---

### Time required
About 50 minutes

This text based activity takes students through basic knowledge about hair and how shampoos and conditioners work. Short pieces of text are interspersed with questions. The purpose here is to build up background knowledge about the topic.

The activity could be done as devised, but could also be used to provide background reading to other sections. For example, in **Shampoo in the shop**, students take part in a role play marketing a shampoo and in **Bad hair day – some hair headaches** explain advice given by a hairdresser to clients. In both cases, background information would be useful.

More information is available in magazines about hair. These are usually not technical in chemical terms, but have a lot of useful details about hair products, how to use them and how they work. See resources list for suggestions.

### Answers

#### Hair    •

1. Our body hair would probably grow much thicker as we adapted to living without extra warm layers on the skin. The hair would grow to take the place of clothes.

2. Different hair types are caused by the amount of grease produced – too much and we get greasy hair, too little and hair is dry.

3. Hair would probably be more healthy if we used less products because then we would not need to wash these off as well.

#### Washing hair – how does shampoo work?

4. Sodium laureth sulfate because it is less irritating.

5. The two surfactants have different actions – one cleanses and the other conditions.

6. This will depend on the products. 'Greasy' hair shampoos will usually have 'lauryl' sulfates, whereas 'frequent wash' will have 'laureth' sulfates. Greasy hair will need stronger detergents and mild shampoos must not irritate skin.

7. Conditioners help to smooth down the cuticles on the hair fibre surfaces.

**Making hair manageable and shiny**

8. Conditioners help to make hair reflect light in a uniform way.

9. Shampoos are usually slightly acidic. This is to help ensure the chemical compounds in hair can bond to each other making smooth fibres.

## *Making shampoo*

*Index 05.08*

Learning objectives
- To find out what goes into a shampoo
- To find out how a shampoo is made
- To understand problems of scaling up a procedure.

Time required
About 30 minutes

This is a very pleasing activity. The quality of the shampoo product is very good, provided the instructions are followed as stated. It is important to measure the amounts accurately. The essential oil is optional – but it adds a good perfume. These oils are very concentrated so only the small amount shown is needed. The herb extract lends a 'natural' quality to the product and some individuality. Students could prepare this at home (it is no more complicated than boiling up frozen vegetables or cooking pasta) to bring in for the shampoo-making lesson.

The main challenge is sourcing the ingredients. They are available from Neal's Yard Remedies (see resource list) whose help is acknowledged in the preparation of this exercise. However, they are also easily obtainable through the many websites for home soap and shampoo makers (see resource list). A suitable container would be a sterile 100 cm$^3$ bottle with screw cap, which will probably cost more than the reagents.

A second point is what to do with the shampoo. It seems a great pity to throw it away just because it has been made in school and therefore has some (unspecified) 'risk' attached. I have never heard of anyone not eating a cake or trifle just because it was made in school. If the the solutions and components have come from a source intended for shampoo making and if the measuring equipment and area used in preparation are scrupulously clean and the bottle sterile, then common sense should prevail - this product is no more 'dangerous' than any other and is probably a lot gentler on the skin and hair than most! However, bear in mind that no preservative has been added, so the product should not be kept longer than three weeks and anyone suffering from skin complaints such as eczema should NOT use the product. **Do NOT be tempted to use ordinary laboratory reagents which may be contaminated.**

Chemical suppliers

Neal's Yard Remedies provided the formulation for the shampoo here and may be prepared to provide small quantities of the chemicals on request, although this cannot be guaranteed, as they are not chemical suppliers. Neal's Yard do sell essential oils, however. Contact information is:
Neal's Yard (Natural Remedies) Ltd
26-34 Ingate Place
Battersea
London SW8 3NS
Telephone: +44 (0) 207 498 1686
Fax: +44 (0) 207 498 2505
website: *www.nealsyardremedies.com* (accessed November 2003)

*www.sourcerer.co.uk* (accessed November 2003) is a directory of chemical suppliers - this is a very useful website for tracking down anything!

*www.chemistrystore.com* (accessed November 2003) is based in the US, but has an excellent range of products – including shampoo specifications already compiled. The recipes are slightly different, however.

*www.tri-esssciences.com/* (accessed November 2003) is also based in the US. This company prides itself on supplying precise needs to schools, students and teachers.

### Answers

1. The shampoo includes a herb extract. This is not found in most commercial shampoos.

2. Ammonium lauryl sulfate – surfactant, polysorbate-20 – emulsifier, essential oil – perfume and the foam-maker is lauryl betaine.

3. Mixing thoroughly to make sure the product is the same all the way through; making sure the oily and watery components mix properly; getting enough herb extract.

## *Results table for shampoo test*

*Index 05.09*

## *Shampoo in the shop*

*Index 05.10*

---

### Learning objectives
- To experience how a shampoo is marketed
- To devise and carry out tests on a shampoo for safety, stability and reliability
- To communicate scientific ideas to the public.

---

### Time required
At least 2 x 60 minutes

This activity shows the background behind putting a cosmetic product on the supermarket shelf. It was designed with the help of a development manager working in the industry. Students should be placed in teams of no more than four. If there are three, then the one should take two posts – the Development Manager's role and one other.

The Development Manager is the nominal leader of the team. S/he should ensure that the others are completing their tasks and be responsible for co-ordinating the launch. This is the role for students who are talented organisers. Artistic students may like the role of Design and Packaging Artist. Although not as scientific as some of the others, the impression the product gives is a clue to its role and function – we are influenced by packaging, like it or not!

Students should carry out tests on their shampoo product. This could be the shampoo made in the experiment **Making shampoo**. The tests should ensure that the product 'marketed' is safe, stable and reliable. In other words, that no one can be harmed by using it, that the product will not decompose or go mouldy and that it does what it says. There is wide scope here for developing investigations and cross-curricular activity with technologists and designers. Also, local connections with anyone connected with marketing or the cosmetics/hair industry may be useful contacts, perhaps especially at the 'launch' lesson.

At the start, set a date for the product 'launch', so the investigations have a time limit. At the launch, products should appear in their packaging, with reports on their stability, safety and reliability. The marketing team should also advertise their product. You may wish to set this up as if you are a major contractor seeking to give a large contract for making shampoo. You will hear the presentations made and make a decision as to which team should get the contract. Good display material will be one additional defined outcome of this activity.

***Person profile – Rahila Begum***

*Index 05.11*

 ***Bad hair day – some hair headaches***

*Index 05.12*

---

**Learning objective**
• To apply chemical knowledge about hair to solving hair problems.

---

**Time required**
About 40 minutes

This activity applies information provided in **Hair and shampoo – the facts** to 'real life' situations. It could be run in one of two ways:
- As a written task done by individuals – each student needs a copy of the sheet and to write their own answers for each person. This could be a suitable homework activity.
- As a discussion task completed in groups – students could discuss the clients in small groups. Their answers could be written on A3 laminated cards using wipe-off markers and presented to the class as a whole.

**Points to look out for:**

**Hayley** – split ends are caused by a damaged cuticle, for example by blow-drying too hot, damaging the cuticle by brushing roughly or too often. They can only be treated by being cut off.

**Tim** – dandruff is a build up of dead cells. We all shed skin cells, but in dandruff the sizes are larger than normal. The causes are described in **Did you know? About dandruff**. The medicated shampoo will help prevent bacterial growth and the herb extracts will improve the condition of the skin, helping it to stop shedding cells.

**Rob** – Too many hair treatments can damage the cuticle. This means the 'roof tiles' get roughened up. Using a silicon-based conditioner will help to smooth the cuticles again. However, it is impossible to truly repair the cuticles, so Rob must wait until his hair grows again.

**Tina** – Her hair is also chemically damaged by a build-up of chemicals getting into the hair fibres and interrupting the bonding. Using a frequent wash shampoo will help to make sure the level of chemical build-up on her hair is minimised. Using a conditioner will help to make sure the cuticles are kept smooth.

## *Did you know? The truth about dandruff*

*Index 05.13*

## *Fringe benefits – investigating shampoos*

*Index 05.14*

Learning objectives
- To decide on data to collect to answer a question
- To decide how to test for certain factors in a shampoo
- To use scientific knowledge in deciding what to test.

Time required
- At least 40-50 minutes to plan an investigation.
- About 60 minutes to carry out an investigation.
- Extra time to write a report.

This activity is based on the kinds of questions people ask about shampoo. Students are encouraged to think about how to investigate these. Background research could be carried out using websites listed in the resources list. The other activities in the unit also provide useful background information.

Points to develop include:
- how to make objective measurements of factors like 'hair shine', 'hair feel', 'hair texture' etc;
- how to test without using animals – or even humans; and
- how consumer testing is done.

In writing a report, the suggestion is to make it accessible for a non-scientist reader. Students can consider the format which would be best for a client to read.

Companies such as Boots the Chemist in the UK have strong commitments to education so may be prepared to reveal information about how they test shampoos.

## *Did you know? Hair today, gone tomorrow*

*Index 05.15*

---

## *Further information, references and source material*

Websites with information about hair include:

*www.keratin.com* (accessed November 2003) this features very well-written, short texts about all aspects of hair, together with scientific references.

*www.Free-Beauty-Tips.com* (accessed November 2003) is a more wide-ranging site, so the hair pages are less scientific and extensive, but the information presented is very useful.

*www.iVillage.com* (accessed November 2003) is a very good source of information on hair, particularly practical solutions to problems. Technical matters are accessed through appropriate links.

**Well-known hairdressers and hair dressing chains also have websites, for example:**

*www.toniandguy.co.uk/start.html* (accessed November 2003) hairdressing chain Toni & Guy

*www.cwlondon.com/* (accessed November 2003) Charles Worthington

*www.nickyclarke.co.uk/* (accessed November 2003) Nicky Clarke

**Magazines focusing on hair include:**

*Hair* published monthly by ipc SouthBank Publishing, widely available in newsagents.

*Cosmopolitan Hair and Beauty* published by National Magazine company, also widely available.

Articles on hair appear regularly in women's and men's magazines, so any of these could be used to provide pictures, other material and background for the activities here.

**References**

D. Fox, *Keep your hair on*, New Scientist 23rd October, 2001, p 26 – 35.

*Hair magazine 7 ways to make your hair shine* December/ January 2002, p 91.

*Neal's Yard Remedies  Natural Health and Body Care* London: Aurum Press 2000.
This is a very beautiful book with lots of good information about all aspects of health care.

R. Winter, *A consumer's dictionary of cosmetic ingredients 5th Edition*  New York: Three Rivers Press, 1999.
This is a comprehensive guide to every ingredient used in making cosmetics.

# Chemistry and diet

# RS•C      Chemistry and diet

**Summary**
The activities reveal more about fat, leading to the consideration of a 'fat pill' as a way of controlling over-eating.

| Resource name | Index | Type | Age range | Topic | Media |
|---|---|---|---|---|---|
| *Fat food* | *06.01* | Paper based exercise – data consideration | 11–16 | Data consideration | |
| *Fat food information table* | *06.02* | Database | 11–16 | Database | |
| *Did you know? About fat* | *06.03* | Information | 11–16 | Fat molecules | |
| *Frying tonight* | *06.04* | Experiment | 11–16 | Making chips | |
| *Key words* | *06.05* | Glossary | 11–16 | Glossary | |
| *Did you know? About taste and smell* | *06.06* | Information | 11–16 | Taste and smell | |
| *Fats, oils and flavour* | *06.07* | Experiment | 14–16 | Unsaturation | |
| *Did you know? More about fat* | *06.08* | Literacy / DART | 14–16 | Saturated and unsaturated fatty acids | |
| *Making a lipid* | *06.09* | Interactive activity | 14–16 | Making a lipid | |
| *Perfect pizza* | *06.10* | Literacy / DART | 11-16 | Food technology | |
| *The 'fat pill'* | *06.11* | Literacy / DART | 11–16 | Diet pills | |
| *Did you know? About obesity* | *06.12* | Information | 11–16 | Obesity | |
| *Case studies* | *06.13* | Group work analysis and discussion | 11–16 | Lifestyle | |

Key:  ⊘ Interactive student activity    ℗ Photocopiable and printable worksheet    ◉ Projectable picture resource

# RS•C

## Chemistry and diet

Issue

Most people eat too much fat. Obesity is an increasing problem in developed countries. Is taking the 'fat pill' the answer to helping people lose weight?

Chemical topics
- Unsaturated and saturated compounds
- Structure of fats and oils
- Reaction of iodine across a double bond
- Food Technology topics.

Scientific enquiry issues
- Scientific developments can cause disputes over use
- Science can respond to practical need.

Activities
- Consideration of data
- Class experiment – cooking chips
- Class experiment – reaction of iodine with unsaturated fats and oils
- Text activity on the perfect pizza
- Text activity on the 'fat pill'
- Analysis and discussion of case studies.

Age range

11–16: see notes. Could also be used in AVCE project work in unit 6.

Notes on using the unit

The background required to fully understand the activities is more appropriate for 14–16 year olds than 11–13 year olds. However, the activities themselves would all be suitable for use with younger children – care needs to be taken with the background information.

The activities could all be used a free-standing tasks in different contexts. Each requires 40–60 minutes of lesson time. If there are large-sized students among those doing these activities, extra sensitivity will be required. Nevertheless, the issue of obesity is a serious one, so addressing the issue of diet and exercise vs. 'quick fix' is important.

**Fat food** is an introductory activity to get students thinking about fat in their diets. They are encouraged to think about foods they eat which contain fat, to compare these against a list and to decide what figures would mean 'low' and 'high' fat.

Next, the chemistry of fats and oils is introduced. In this activity, students usefats and oils to see if there is any link between the flavour and the type of molecules present. They investigate the fats to see the extent of unsaturation in the molecules.

In **Perfect pizza**, students begin to apply the information to new situations. This activity describes scientists' search for low fat cheese to make a pizza with the same qualities as a normal pizza. **The 'fat pill'** introduces the idea of the 'fat pill', a drug which helps prevent absorption of 30% of ingested fat. Students analyse the results of the clinical trial. Finally, three case studies of very overweight people are presented for whom students are asked to suggest treatment plans.

Food chemistry is a very big area of science – students may wish to consider careers as dieticians or nutritionists and for both study of post-16 chemistry is a pre-requisite. These activities introduce this aspect of chemistry in a topical way likely to generate students' interest.

The activities could be given to students under the guise of investigating fat in the diet working as a dietician for a  health care trust. The case studies could be presented first, with the idea that in order to recommend treatment plans, students would first need to analyse the diets of the patients, work out more about fat chemistry and finally recommend a programme.

## *Fat food*

*Index 06.01*

---

### Learning objectives
- Food items contain different amounts of fat
- Deciding what is 'low' and 'high' fat is arbitrary
- People have different needs for fat in the diet
- Fats and oils are lipids with a specific chemical structure
- Food labelling can be deceptive.

---

### Time required
About 40 minutes

The list of foods is based on a fairly typical western-style diet. As an alternative, students could be sent away to find foods which are low and high in fat. Food labels in the UK have to carry nutritional information – this is a good opportunity to help students make sense of this. Fat is often quoted as grammes per 100 g. This is a good way of expressing fat as a percentage of mass, so adding to the data in the table.

The first step could also be done as a class discussion. For example, students could be asked which foods they think of as 'low' and 'high' in fat, with reasons. Most reasons will be based on labelling, what they have been told from sources such as advertising. Few students will have read detailed nutritional content labels.

Typical 'low' fat foods may include: yoghurt, skimmed milk, any food labelled 'less fat', fruit, chicken, rice, pasta, bread, vegetables.

Typical 'high' fat foods may include: pizza, cheese, chips, cream, pastry, butter, meat, chocolate, curry.

---

### Answers

1. Students can discuss what should be regarded as sensible figures for both 'high' and 'low'. Information from food labels may help – do manufacturers work to the same figures? That is, can you find 'low' fat foods with a wide range of fat percentages? As a guide, 'low' could be considered as 15% or less and 'high' 65% or more.

2. The answer will depend on the first part of the activity.

3. Foods for overweight people include: bread, chicken, egg white, fish, milk and turkey.

4. Foods for people needing to put on weight include: butter, bacon, cheese, clotted cream, brazil nuts and lamb chops.

5. Food groups missing are fruit and vegetables, also pulses. Other foods like pizza, chips and pasta are also missing.

6. Not necessarily. A vegetarian diet can also be high in fat – nuts and dairy products can be eaten.

7. 85% fat free means that fat is contained in 15% of the product. This does not mean that there is not much fat – the part which contains the fat could be 100% fat. 'Low fat' is a meaningless term as there is no figure which is regarded as 'low'. The terms are used by the food industry to make us think we are eating less fat.

## *Fat food information table*

*Index 06.02*

## *Did you know? About fat*

*Index 06.03*

## *Frying tonight*

*Index 06.04*

Learning objectives
- To maintain fair conditions for a food test
- To compare results subjectively rather than objectively using measurement
- To realise that cooking can be scientific
- That fats and oils have different properties and flavours.

Time required
If each group tests one fat or oil, about 30 minutes.
Each frying test requires about 20 minutes, so add 20 minutes extra for every bag of chips to be made.

**Important note**
This activity should only be done if Food Technology facilities are available and teachers are totally confident in the behaviour of their students. Otherwise it could be demonstrated in the Food Technology room. Teachers should refer to guidelines prepared by CLEAPSS (specifically *Model Risk assessments for D&T – reference no. 3.019 – Frying and Grilling Food*). Teachers should also be fully aware how to deal correctly with fat and oil fires.

The key here is to ensure the potato pieces are dry before placing in the hot fat or oil. Also, the fat or oil must not be allowed to heat to smoke point, or there is a risk of fire. However, the quantities involved are very small, so this is minimised as far as possible.

In tests, olive oil seemed to be a favourite. Students can also be encouraged to think about the problems associated with developing a taste scale – we are used to making objective measurements using scales and equipment designed for this, so assessing taste requires different skills. Food scientists at major companies are trained in taste analysis using a series of compounds which they have to 'learn' like a language. This is a way of standardising taste and smell.

Answers

1–2 Depend on the students.

3. Information about the fat or oil can be found by doing a web search, or by reading **Did you know? About fat** and **Did you know? More about fat**.

4. The variety of potato and temperature of the oil may also affect the flavour of the chips.

5. Food scientists break down tastes into the component compounds, then mix them together to create specific flavours. Taste is in fact mainly smell, so smelling compounds is often the main component in flavour. By producing the component compounds, a range of smells is available. The scientists learn these like a language, so they can recognise the smells by the compounds they contain.

  ## *Key words*

*Index 06.05*

## *Did you know? About taste and smell*

*Index 06.06*

 ## *Fats, oils and flavour*

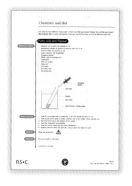 

*Index 06.07*

Learning objectives
- That iodine reacts with unsaturated fatty acid chains in lipid molecules
- Addition reactions occur across a carbon-carbon double bond
- Fats and oils have different component fatty acid chains.

Time required
About 40 minutes, depending on the number of fats and oils tested.

The activity is a very simple way of introducing students to unsaturation in fats and oils. The test with iodine could be extended to calculation of iodine number, but this is outside the scope of this resource. There is, however, the possibility of using this activity as the basis for an investigation into the chemistry of fats and oils.

Only *small quantities* of the fats and oils and iodine are required. The iodine will first appear as dark brown droplets in the liquid. Stirring causes these to disperse. It is possible to watch as the iodine colour 'disappears' into the oil - the tiny droplets seem to simply vanish as the reaction takes place. The final colour may be close to the original oil, but it is difficult to judge precisely the exact amount of iodine to be taken up, so there will be excess. The final colour, so the point when timing should stop, is when a pale pink colouration is present and no more droplets of iodine can be seen. Warming the reaction mixture is helpful – in tests the water was 60–70 °C. Hot tap water would suffice. Results could be duplicated to ensure consistency.

The information table provided with **Did you know? More about fat** can be used to extend discussion of the results of this activity, as the types of fatty acid chains present in the oils and fats can be discussed and the timings compared to the degree of unsaturation.

Apparatus and equipment (per group)
- 5–6 test-tubes
- Test-tube rack

- Supply of clean plastic dropping pipettes, one for each oil
- White card for background
- 250 cm$^3$ beaker
- About 150 cm$^3$ water at 60–70 °C
- Glass rod
- Eye protection for each student
- 5 cm$^3$ of each oil or melted fat to be tested
- Pipette or measuring cylinder for each oil
- 3 drops 2% iodine (about 0.1 M) in 0.25 M potassium iodide solution.

Notes on the requirements

- Students could supply their own oils bringing small quantities from home (about 1 tablespoon per test)
- The oils can be disposed of down the sink with water
- Melted fats should be poured into a beaker to solidify before disposal in the refuse
- Provide a bowl of hot, soapy water in which students can place the used oily test-tubes.

Results

| Oil | Time minutes | Comments |
| --- | --- | --- |
| Extra virgin olive | 3.5 | Could be compared with non-extra virgin. Quite a dark coloured oil, so harder to determine end point. |
| Peanut | 3.0 | A very light coloured oil. End point easy to see. |
| Cod liver | 1.5 | The short times for these oils makes the point about unsaturation very clearly. |
| Soya | 1.5 | |
| Sunflower | 1.0 | |

Saturated fats such as butter would not change after 5 minutes.

Answers

1. These results suggest that sunflower oil has the highest proportion of unsaturated fatty acids.

2. Extra virgin olive oil has the least unsaturated fatty acids.

3. Unsaturated fatty acid molecules break up more easily and they help to prevent heart disease. In this case, the olive oil had the most saturated fat and the best flavour.

  ## *Did you know? More about fat*

*Index 06.08*

Answers

1. (a) Cocoa butter and beef fat  (b) Corn oil, fish oil and olive oil.

2. Those containing unsaturated fatty acids.

## *Making a lipid*   Index 06.09

This activity animates the formation of a lipid and shows ow a small molecule is thrown out. It is a condensation reaction.

## *Perfect pizza*

Index 06.10

---

Learning objectives
• To consider how scientists solve a 'food' problem
• To apply differences between low and high fat food to diet
• Food components have different amounts of fat.

---

Time required
About 25 minutes

This short text shows how scientists are trying to produce low fat versions of our favourite foods. Here, low fat cheese is being used to produce a pizza. Scientists discovered how cheese melts – that an oil forms on the surface and that this 'burns' to give the burnt cheese taste on top, protecting the melted cheese underneath. The oils come from within the cheese.

When low fat cheese is used, there is much less oil within the cheese, so in cooking the pizza turns out dry and the cheese crumbly. To compensate for the lack of oil, scientists realised that spraying on a small amount of extra fat could give the same effect as for the original mozzarella.

The investigations are ways of taking the work further. Supermarkets sell 'pizza cheese' ready grated in packets – these could be compared with a solid block of mozzarella for fat content as a starting point.

Note – if any investigation involves tasting, this should only be carried out in Food Technology facilities.

Answers

1. A lot of people eat pizza, but it is a high-fat food, so can add to dietary problems.

2. In low fat cheese egg white or soya protein replaces fat (see **Fat food information table** for fat levels).

3. Low fat cheeses melt to dryness and form a hard skin on top. They do not melt like 'normal' cheeses.

4. Scientists made observations about how mozzarella cheese melts. They saw the oils coming up to the surface. They realised that spraying a low fat cheese with a small amount of extra oil would give the same effect.

5. They should test the oils for flavour. Also, they should test the cheeses for flavour. It is OK doing this, but they need to mimic the tastes as well. Scientists could be producing a low fat but tasteless product.

## *The 'fat pill'*

*Index 06.11*

---

### Learning objectives
- Drugs can be taken to help people lose weight
- Xenical is a drug which inhibits lipases
- A drug can stop fat being taken into the body
- A 'fat pill' alone will not bring much weight loss.

---

### Time required
About 25 minutes

Xenical (also called Orlistat) has undergone clinical trial in the USA and is now licensed for use there. Xenical can be purchased over the internet. The drug was isolated from soil bacteria and inhibits lipases which break down fat molecules. This can reduce fat intake through the wall of the intestines by up to 30%. There are side effects – the extra fat has to pass through the rest of the digestive system and can cause flatulence and constipation.

Xenical acts in contrast to other diet pills which actively suppress appetite by affecting neurotransmitters in the brain. These are perhaps more dangerous, as overdoses of these can influence the entire intake of food. Xenical represents a 'safer' way of controlling the intake of fat, a major contributor to obesity.

However, taking any drug to reduce weight must be considered a drastic step to take, reserved only for those who are clinically obese or have other extremely good reasons for needing to lose weight other than by dietary and exercise changes. The activity can be used in the context of health education to support work on prevention of anorexia, the need to maintain a healthy weight and body image.

The activity is also an introduction to biochemistry, picking up on the role of enzymes in digestion. This could be extended by additional research on lipases, perhaps with experiments.

It should be stressed to students that a little knowledge is a dangerous thing – this is only an introduction to a serious topic and that medical expertise is needed before making judgements about their own situations.

Answers

1. Most fat pills work by reducing appetite – they work on the brain.

2. Xenical works on the gut and is not ingested – the drug inhibits enzymes which break down fat molecules.

3. There is some evidence suggesting that Xenical does help, but the differences between Xenical and non-Xenical patients in the clinical trial were quite small.

4. No, simply taking Xenical would not really allow a person to go on eating. Their fat intake would probably be too high for good health anyway.

5. No, taking any drug to change how food is metabolised is a serious step to take. People who are clinically obese could be prescribed Xenical, but not someone who is just a bit overweight.

## *Did you know? About obesity*

*Index 06.12*

## *Case studies*

*Index 06.13*

Learning objectives
- To apply knowledge about fat in the diet to three 'cases'
- To consider how illness and other factors affect body mass
- To contrast diet and exercise vs drug treatment for obesity.

Time required
About 30 minutes

This is a discussion exercise which draws together the themes presented in the activities. Students are invited to consider the cases of three fictional people who are all severely overweight, but for very different reasons. The point here is to apply knowledge about diet and exercise to develop a simple treatment plan to help the patients lose weight.

Students can also discuss the role of the fat pill in treating obesity. If one of the three patients were to get this drug, who would it be? Issues to consider include why the patients became fat, their motivations for change, what their life situations are and their possible life spans. Two of the three, 'Leanne' and 'Mark' could be argued to have 'good' reasons for being overweight, but assessing their two situations should provoke discussion.

BMI values are:

Leanne    $85.3 / (1.68)^2 = 30.2$
Fiona     $78.8 / (1.57)^2 = 32.0$
Mark      $95.3 / (1.75)^2 = 31.1$

All three are defined as obese. The best candidate to get Xenical is Leanne, as some evidence indicates that this kind of treatment is most beneficial to diabetics.

See *www.geocities.com/nutriflip/Nutrients/Lipase.html* (accessed November 2003) for more information.

## Further information, references and source material

Many websites offer information about the topics in this unit. Those checked on dieting were responsible, indicating the need to seek medical advice before starting on a diet or weight loss plan. Some have extremely sad personal stories written by obese people, which can be powerful to read. Diet pills and Xenical can be bought easily over the internet, so there is a need to educate on this topic.

Websites with information about fat, Xenical and diet include:

*www.uen.org* (accessed November 2003) includes a lot of activities on food science.

*www.obesity.org* (accessed November 2003) is the site for the American Obesity Association. The USA has one of the highest percentages of obese people in the developed countries. This extensive site has a great deal of information on the topic, but everything is in US units.

The BBC's website at *http://news.bbc.co.uk* (accessed November 2003) has nearly 300 articles related to obesity. These show the background to the development of Xenical and other experiments related to fat pills, genetic tendencies towards obesity and data on obesity. Type 'Obesity' in the 'Search' facility to get the list.

*www.xenical-dietpill.com* (accessed November 2003) has information about Xenical.

More general websites with information about health and body mass include:
*www.hebs.scot.nhs.uk* (accessed November 2003) is the Health Education Board of Scotland. A very good report on Childhood Weight Management is available with some useful data on overweight children.

### Articles and other information

BBC *Brits are European heavyweights* *http://news.bbc.co.uk/1/hi/health/2220923.stm* (accessed November 2003) article showing that British people are among the heaviest in Europe – British men have an average mass of 79.75 kg and women 66.7 kg.

Bovsun, Mara *Fat blocking pill aids in weight loss* www.applesforhealth.com/fatblock1.html (accessed November 20032)

Claasen Jojanneke, *On the scent of taste*, Baarn, Netherlands: Tirion Publishing 1994.

C. Dierks, *In search of tasty low-fat pizza* http://www.peregrine-pub.com/news/pizza.html (accessed November 2003)

The Economist, *The fat of the lands*, 23rd February, 2002 p 115–116.

T. Gura, *New Scientist*, 2001, 172, 56.

The Guardian, *Land of the fat*, 2nd May, 2002. *www.guardian.co.uk/medicine/story/0,11381,708413,00.html* (accessed November 2003)

M. Hall, *Diabetes*, London: Association of the British Pharmaceutical Industry 1999.

M. Heil, *Better Pizza through chemistry* Discover 20 no 9 (September 1999) *http://www.discover.com/sep_99/pizza.html* (accessed November 2003)

K. Kleiner, *New Scientist*, 2002, 173, 9.

R. Oesch, *Fat burner pill to help diabetics lose weight* 8th April, 2002 at *www.wdef.com/health/MGBBLZNTSZC.html* (accessed November 2003)

G. Watts, *New Scientist*, 2002, 173, 29.

# Vitamins

# RS•C    Vitamins

**Summary**
This material considers vitamins and in particular focuses on vitamin C. Discussions about the role of vitamin C in the prevention of colds and the need for vitamin supplements are presented using scientific evidence.

| Resource name | Index | Type | Age range | Topic | Media |
|---|---|---|---|---|---|
| *Vitamin C* | *07.01* | Paper based literacy / DART and questions | 11–14 | Vitamin C, scurvy and scientific method | P |
| *Key words* | *07.02* | Glossary | 11–16 | Glossary | ● P |
| *Did you know? About vitamin C* | *07.03* | Information | 11–16 | Sources of vitamin C | ● P |
| *A scurvy solution: testing for vitamin C* | *07.04* | Class experiment and questions | 11–16 | Testing for vitamin C | P |
| *Did you know? About Casimir Funk* | *07.05* | Information | 11–16 | The discoverer of vitamins | ● P |
| *Catching a cold?* | *07.06* | Literacy and data analysis questions.  Interactive activity | 11–16 | Colds | ● P |
| *Did you know? About Linus Pauling* | *07.07* | Information | 11–16 | Linus Pauling | ● P |
| *A cold survey* | *07.08* | Class survey / investigation Questions and presentation exercise | 11–16 | Data collection and analysis | P |
| *A cold cure? Advice please!* | *07.09* | Class paper based design exercise | 11–16 | Presentation exercise | P |
| *The pill thrill: are vitamins a waste of money?* | *07.10* | Discussion and presentation exercise | 11–16 | Presentation exercise | P |

Key: ● Interactive student activity    P Photocopiable and printable worksheet    ◉ Projectable picture resource

# RS•C  Vitamins

## Issue

Taking vitamin C is claimed to prevent colds. Vitamin supplements are sold as being vital to good health. Students are invited to take a critical look at data, collect their own and decide if taking vitamin pills is really necessary.

## Chemical topics

- Vitamin C can be found in many types of fruit juice
- Titration of vitamin C using an oxidation reaction
- Vitamin C concentration can be analysed.

## Scientific enquiry issues

- Presenting scientific ideas
- Fair testing
- Judging uncertainty in observations and measurements
- Considering the reliability of scientific evidence
- Communicating data using diagrams, charts, graphs and tables
- Using evidence to support a conclusion or interpretation.

## Notes on using the resource

In **Vitamin C** students read two texts written in old-style English. These introduce the role of vitamin C in the body and the idea of a clinical trial.

**A scurvy solution: testing for vitamin C** describes a class experiment to determine the relative amounts of vitamin C in different fruit juices. The technique is a simple titration using iodine as the oxidising agent of ascorbic acid. A wide range of different juices can be tested. Extensions of the experiment to determine concentrations of vitamin C and a range of investigations are described.

The following three activities can be used independently. However, to make most sense, teachers might consider using them sequentially. In **Catching a cold?** students are presented with data used by Linus Pauling in making his claim that vitamin C prevents colds. In **A cold survey** students carry out their own 'cold survey' and analyse and present their data. Finally, in **A cold cure? Advice please!** students are invited to consider two reports written more recently and put this together with their own evidence to write a health advice leaflet. The reports are chosen deliberately to show two different views on vitamin C.

In **The pill thrill: are vitamin pills a waste of money?**, four different views about taking vitamin pills are presented. Students are invited to discuss these in groups and to rank them in order of agreement. The objective is to decide if taking vitamin pills is a waste of money or not. Teachers could encourage students to use evidence gathered from other activities, if these have been completed first.

 ## *Vitamin C*

*Index 07.01*

109

# Vitamins

Learning objectives
- To find out why the body needs vitamin C
- To find out the characteristics of a clinical trial.

Time required
About 40 minutes

This activity is designed to introduce the role of vitamin C in the body. The first text describes the death of a sailor on board a ship in the 18th century. The key points to draw out here are that the sailor was very weak overall, and that he was probably bleeding inside as well as from old wounds on his skin. His gums had shrunk so that his teeth could not be held in place any longer. These symptoms are now known to be due to vitamin C's role in the production of collagen, a key protein involved in the body's skin and bone structure. Some students with lower reading abilities may need help with certain phrases in this text.

The second text is a classic description of the first ever clinical trial. The text has been amended and a glossary of terms is provided to help students establish meanings for the more old-fashioned words. Teachers may wish to discuss why it took so long for the Navy to change their practice and give lemons as part of the sailor's diet – and does the same kind of thing happen today? What is the role of scientific evidence in helping non-scientists make changes?

Other routes to take with this activity may include inviting students to bring in labels from different fruit juices or other foods showing the amounts of vitamins present, such as cereal packets. This creates the possibility of discussing the role of different vitamins in the body in addition to vitamin C.

Notice that in this activity the link between cold prevention and taking vitamin C is not mentioned – in the early stages vitamin C was recognised first as a contribution to overall health rather than linked with the common cold. Today a range of studies indicate the place of vitamin C in prevention and also contribution to disease (see resources list), but the main perception is the link with the common cold.

## Answers

1. Weakness, opening up of old wounds, loose teeth, bleeding gums, bleeding around hairs on arms.

2. Their diet was restricted to the food carried on board ship. If this did not have enough vitamin C to keep them healthy, they would run out of vitamin C and then eventually get scurvy. Sea voyages often lasted for several years so there was lots of time for scurvy to develop.

3. Their diet included water-based porridge, sugar, meat, biscuits, puddings, other grains like rice and barley, dried fruits and wine.

4. In a way the sailors' diet could be described as healthy, as it included a range of different foods. There is no sign that they starved. But from what we know today, the diet was not healthy because it did not include fresh fruit. The sailors and Navy did not know this at the time.

5. The treatments were:-
drinking about a litre (1 dm$^3$) of cider, drinking sulfuric acid, drinking vinegar, sea water, taking oranges and lemons and eating a paste made from garlic, mustard seed and other substances. The treatments were usually taken on an empty stomach.
He probably chose these because these were known to him, perhaps from medical training, discussions with colleagues or his own ideas. Some of the treatments probably came from his medicine chest. It is not clear why he chose to give the men fresh fruit.

6. He tried different treatments to see which had the best effects.

7. He gave the treatment to two sailors in case one died, and also to test if the treatment worked on more than one person. If it worked on one patient only then probably it was not a good treatment.

Testing on two reduced the chance factor.

8.   He tried to give his treatment to similar cases, living in same conditions, on roughly the same diet. Improvements would be:
     • Increasing the numbers of sailors in the groups
     • Reducing the number of treatments and testing two or three at a time with larger groups
     • Keeping the men apart from each other in case other diseases and infections caused some of the symptoms
     • Asking someone else to give the treatments so he did not know exactly who got which – then comparing the results. This would be making the trial a 'blind' trial.

9.   Yes, Dr. Lind's conclusion was correct. Cider would also probably be the next best treatment, but the level of vitamin C would be much lower than that in oranges and lemons.

Note that oranges and lemons were not added to sailors' diets for another fifty years – partly because Dr. Lind himself was not entirely convinced of the results of his work!

Teachers might also like to use further activities on scurvy as described in D. Warren, *The nature of science*, London: Royal Society of Chemistry, 2001, in the chapter entitled Scurvy – the mystery disease, page 34.

## *Key words*

*Index 07.02*

## *Did you know? About vitamin C*

*Index 07.03*

# A scurvy solution: testing for vitamin C

*Index 07.04*

---

## Learning objectives
- To find out differences between the amounts of vitamin C in a range of fruits
- To carry out a simple titration
- To use an oxidation reaction in a titration.

---

## Time required
50–60 minutes

This is a pleasing experiment which yields reliable results. The titration is carried by counting the number of drops of fruit juice/vitamin C solution needed to reduce a fixed amount of iodine. The iodine solution is prepared in a test-tube with starch indicator. The fruit juice is added dropwise until the blue-black coloration is no longer present. In practice, the indicator will not become completely colourless, but a grey-white colour will suffice as the endpoint.

The basic titration can be extended, for example:
- Preparing a standard solution of vitamin C from a commercial tablet yielding a known mass. Dilution of the standard solution to a range of concentrations enables preparation of a standard curve, as each diluted solution could be tested against the iodine. The standard curve could be used to read off the concentrations of the vitamin in the fruit juices.
- Placing different fruit juices in burettes around the lab. Students would be able to titrate the juices more accurately and test a range.
- A range of investigations is also possible based on this experiment. Suggestions for these are given in the students' sheets.

## Technical requirements

Before the lesson these solutions are needed:
- A solution of 0.05M iodine in potassium iodide prepared from 1.27 g iodine and 1.5 g potassium iodide made up to 100 cm$^3$ with distilled water. See CLEAPSS recipe cards for full procedure
- 0.1% starch solution
- Vitamin C solution prepared by dissolving a 100 mg tablet in 100 cm$^3$ water (the solution is then 1 mg/cm$^3$).

## Apparatus and equipment (per pair of students)

- 10 cm$^3$ vitamin C solution (1 mg/cm$^3$)
- 1 drop per test iodine solution
- 10 cm$^3$ starch solution
- About 50 cm$^3$ water
- Fruit juices to test –have the packaging available if the fruit is not fresh
- Droppers or plastic pipettes, one for each solution and fruit juice
- Test-tubes, a clean one for each test
- Test-tube rack
- 10 cm x 10 cm piece of white paper or card for background
- Eye protection.

The class will also need a range of fruits to test. These are best bought fresh on the day of the experiment. Small amounts of juice can easily be squeezed out by hand, but using a juicer would help get enough for a class. Alternatively, students could bring in their own juice squeezed in advance in a secure container. A comparison could be made between packaged fruit juices and fresh fruit.

Table 1 gives the quantities of vitamin C in a variety of different fruits. This can be used as a guide to the range of fruit that could be tested.

| Fruit | mg vitamin C /100 g | mg vitamin C per fruit or slice |
|---|---|---|
| Apple | 6 | 8 |
| Apricot | 10 | 4 |
| Avocado | 8 | 16 |
| Banana | 9 | 11 |
| Baobab | 150 – 500 | 100 |
| Breadfruit | 29 | 28 |
| Blackberry | 6 | 0.60 |
| Blackcurrant | 155 – 215 | 1.5 – 2 |
| Blueberry | 1.3 – 16-4 | no data |
| Camu camu | 2 700 | no data |
| Fig | 2 | 1 |
| Grape | 11 | 0.60 |
| Grapefruit | 34 | 44 |
| Guava | 183 | 165 |
| Jujube | 500 | no data |
| Kiwi fruit, green | 98 | 74 |
| Kiwi fruit, yellow | 140 | 140 |
| Lemon juice (one wedge) | 46 | 3 |
| Lime juice (one wedge) | 29 | 1 |
| Lychee | 72 | 7 |
| Mango | 28 | 57 |
| Melon, honeydew | 25 | 20 |
| Melon, cantaloupe | 42 | 29 |
| Orange | 53 | 70 |
| Papaya | 62 | 47 |
| Pawpaw | 14 | 28 |
| Passion fruit | 30 | 5 |
| Peach | 7 | 6 |
| Peach, canned | 3 | 3 |
| Pear | 4 | 7 |
| Pineapple | 15 | 13 |
| Plum | 10 | 6 |
| Raspberry | 25 | 1 |
| Redcurrant | 70 | 0.70 |
| Rosehip | 1150 | 45 |
| Strawberry | 57 | 7 |
| Tangerine/mandarin | 31 | 26 |
| Tomato | 19 | 23 |
| Watermelon | 10 | 27 |

*Table 1 Vitamin C quantities in a variety of fruits*
(Reproduced with permission from *www.naturalhub.com*).
*www.naturalhub.com/natural_food_guide_fruit_vitamin_c.htm* (accessed June 2004)

Results - example

| Fruit juice / vitamin tablet | Number of drops to reach end-point |
| --- | --- |
| 1000 mg vitamin C tablet in 100 cm$^3$ | 15 |
| lemon juice | 30 |
| Lime | 35 |
| Orange juice from day-opened carton | 40 |
| Kiwi | 15 |
| Melon | 20 |
| Mango | 30 |

Students may need help to realise that the lowest figures mean the highest concentrations of vitamin C and vice versa.

Answers

1. So you can see the endpoint.

2., 3., 4., 5. The answers will depend on the fruit juices tested.

Investigations

These are suggested for extension of the basic experiment. The questions are of different levels of complexity.

Answers

1. The answers to these questions will depend on the fruit juices tested – answers here are given based on the results table above. The data show that the fruit with the most vitamin C is the kiwi fruit. The samples with the least vitamin C are the orange juice from the carton and the lime. The smaller the number of drops the greater the vitamin C content. This is because more of the iodine has reacted with the vitamin C.

2. The data suggest that kiwi fruits and melons would be good sources of vitamin C. They could also take vitamin C tablets.

3. Advise the Navy: This is a good opportunity for some cross-curricular work – students may investigate the life at sea in the 18$^{th}$ century, the style of writing used at the time, more about Dr. Lind and his work. Websites to help with this are suggested on page 120.

## *Did you know? About Casimir Funk*

*Index 07.05*

# RS•C

## Vitamins

## Catching a cold?

*Index 07.06*

---

### Learning objectives
- To analyse data to establish understanding of a scientific issue
- To be critical of data presented as making a specific claim
- Fair testing
- To consider how the work of one scientist can influence our thinking.

---

### Time required
About 40 minutes

The activity uses some data collected during a study undertaken in 1961 and used by Linus Pauling in his book, *Vitamin C and the Common Cold*, published in 1970. The data are adapted and changed slightly to suit the younger audience.

The objective is to encourage students to look critically at the data – this is good training for reading newspaper reports of scientific information. The data could be interpreted as showing unequivocally that vitamin C helps to prevent colds, but try to get students to look behind the first impressions. For example, there is little information about how the two groups of skiers lived. Several possible other reasons could be that the vitamin C group could have been living at much closer quarters than the placebo group, so were more susceptible to infection; the group could have eaten a bad meal which made a number ill; several people in the vitamin C group could have been very significantly ill for most of the twelve days of the study, skewing the figures.

Students are invited to draw some data from the study and carry out a calculation to show the decrease. The questions try to draw out their critical faculties.

### Answers

1. a) 279
   b) 12 days in total
   c) Vitamin C and placebo
   d) 140 in the placebo group and 139 in the vitamin C group
   e) Yes.

2. a) Vitamin C 17 , Placebo 31
   b) Vitamin C 42 , Placebo 119
   c) Vitamin C 31 , Placebo 80

3. a) 51.61%
   b) 34.45%
   c) 37.50 %

4. The skiers who took vitamin C had fewer colds, symptoms and numbers of days illness than the placebo group. The decreases are quite large numbers, even though the number of skiers in each group was almost the same.

5. Possible answers are:

a) Symptoms not due to colds - tonsilitis, earache, aches in the muscles, headaches, pain in the stomach, vomiting, diarrhoea and general body weakness.
Tonsilitis is due to a bacterial infection, not the cold virus; earaches could be caused by all sorts of reasons, eg altitude, not wearing a hat; aches in the muscles could be caused by skiing/other physical activity; stomach pains, vomiting and diarrhoea are not symptoms of the cold; someone could feel weak due to tiredness.

b) Differences – amount of time spent outdoors; amount of contact with other people; diet – one group could have eaten a bad meal; living conditions – one group may have been living much higher up the mountain in colder weather, or much closer together. Also, a few people in the vitamin C group could have had a very bad illness lasting for the days of the study, which skews the figures downwards.

c) Information on the doses is not given; or details on what is present in the placebo. These could matter because the placebo may have contributed to ill health in some non-intentional way; the dose of vitamin C may be very high and not suitable for use as a 'drug' in normal circumstances for some people.

6. The answer to this will depend on the students. It is important to collect the reasons for their choices.

## *Did you know? About Linus Pauling*

*Index 07.07*

## *A cold survey*

*Index 07.08*

Learning objectives
- To prepare and carry out a survey
- To analyse survey data
- To consider if people may be influenced by the thinking that vitamin C prevents colds
- To present survey data in a way which makes the key points clear.

Time required
About 1 hour to survey a class of about 30 and to carry out preliminary analysis as a class.
2–3 hours if data collected from family and others at home are to be analysed.

This activity invites students to collect their own data which indicates if people think vitamin C prevents colds. A survey sheet is provided, but students could make their own. If relatively little time is available, students could complete this only within the class. Alternatively, the survey instrument could be taken home for completion among students' families. A third possibility is to carry out a survey with members of the public in a supermarket or local shopping centre. The survey sheet can be used either as a guide and responses entered on the summary sheet for each respondent, or could be completed for each individual in the survey for data to be entered later.

Students usually enjoy collecting this kind of data very much, so it may be worth spending some time organising data collection among members of the public. An important part of the preparation and carrying out of a survey involving collecting personal information is to establish an ethical code of practice – names of individuals should not be entered into a computer and stored against the data, and should be recorded only with their permission. Respondents should also be told their data will be used only for research purposes. Students could be invited to suggest other ethical practices.

Answers

The questions take students through the data in a systematic way, encouraging them to combine answers to get an overall picture in some cases.

The answers will depend on the responses given in the survey.

Note that research studies suggest that taking high doses of a type of vitamin C called 'ester-vitamin C' helps to shorten colds and reduces the severity of symptoms. There is no conclusive evidence that vitamin C actually *prevents* people from getting colds. This is an important difference.

## *A cold cure? Advice please!*

*Index 07.09*

Learning objectives
- To consider how more recent reports on vitamin C influences our understanding
- To present scientific information in an understandable way for the public.

Time required
About 1 hour – this activity could be done as homework following the cold survey or other activities on scurvy.

Research on the role of vitamin C has continued since Linus Pauling's book was published. Many studies have been done; the results of many contradict each other. One possible reason for this is that the more recent reports suggest that an 'Ester' form of vitamin C is effective at reducing the length of colds. Earlier studies may have included ester C without realising it, which others did not, hence some showed positive results and others negative ones.

Many studies have explored the roles of vitamin C in a wide range of other serious diseases, including cancer, arthritis and hardening of the arteries. The work suggested here could be used as a basis for taking the study of vitamin C further. Resource materials are suggested at the end of the unit.

Also, research since Linus Pauling's time has investigated the immune system, which is now much better understood and the roles of other substances in preventing and treating colds. This means that the quality of data collected and the background information used in the studies is now much more reliable than that used by Linus Pauling.

Finally, our housing has now improved and lifestyles have changed – many people in western countries now live in centrally heated homes, drive cars and travel to work on crowded trains! These will all have some effect on the way viruses are transmitted between people and the level of resistance to infection shown.

This activity gives students the opportunity to read brief details of two studies carried out 30 and 31 years since Linus Pauling's book was published. The idea is to produce leaflets which present the information in a format suitable for the general public to read – the sort distributed through doctors' surgeries. The best ones could be sent to the local area health authority for comment!

## *The pill thrill: are vitamins a waste of money?*

*Index 07.10*

Learning objectives
- To consider arguments for and against taking vitamin pills
- To discuss arguments in a group and agree on one answer.

Time required
About 30 minutes

Vitamin pills are sold in many outlets – in the UK some companies specialise in selling these and promote them heavily as 'good for you', promoting health and wellbeing. This activity invites students to consider arguments in favour and against taking them, in the light of a major study published in 2002 suggesting that vitamin pills generated no overall beneficial effect on health.

Once students have agreed on their rankings of the points made, the discussion could continue to establish what does contribute to good health? Points to consider are eating a healthy diet, including significant amounts of fresh fruit and vegetables every day; exercising regularly and not smoking. Although these are quite boring, try to get students to see how we are influenced by 'quick fixes' such as taking vitamin pills – in reality these may not help people because their lifestyles are so

RS•C

unhealthy that taking vitamins will not help much!

In undertaking this discussion, students should be able to consider if they need to change lifestyle, the role of vitamin C in cold prevention and know which fruits to take to help keep their vitamin C levels topped up.

## *Further information, references and source material*

Health websites have information relating to this topic. The sites seem to do their best to be reliable – information can be corroborated elsewhere.

News reports about the role of vitamins and vitamin research can be found on all the major news sites. The sites for the BBC and CNN are given.

*www.bbc.co.uk/* (accessed November 2003)
The BBC site has a health section which includes many reports on vitamin C and vitamins generally.

*www.cnn.com* (accessed November 2003)

*www.everybody.co.nz* (accessed November 2003) is a health website from New Zealand. New Zealand has a more open way of trading health and drug products than the UK. This is reflected in the content of the website.

*www.webmd.com* (accessed November 2003) is one of many 'web-doctor' sites. This one is American.

*www.naturalhub.com* (accessed November 2003) has good information about a wide range of health issues, focusing especially on diet. The page 'about fruit' gives links to vitamin C information, including comparing the vitamin C content of different species of kiwi fruit! The site also has links to other websites with good information about vitamin C.

Information about Linus Pauling can be found at these websites:

*www.nobel.se/chemistry/* (accessed November 2003) is the Nobel website.

*http://lpi.oregonstate.edu/* (accessed November 2003) is the website for the Linus Pauling Institute based at the University of Oregon, USA. The Institute carries out research into the role of micronutrients and vitamins in maintaining health.

*www.paulingexhibit.org/* (accessed November 2003) is the website for a travelling exhibit about Linus Pauling's life.

*http://osulibrary.orst.edu/specialcollections/rnb/* (accessed November 2003) is the website with the full collection of Linus Pauling's research notebooks in digitized format. This can be searched for information on all Linus Pauling's research topics, including Vitamin C.

James Lind's 'A Treatis of the Scurvy' can be found at:
*www.people.virginia.edu/~rjh9u/scurvy.html* (accessed November 2003)

Information about vitamin C and scurvy can be found on several sites including:
*www.people.virginia.edu/~rjh9u/vitac.html* (accessed November 2003)

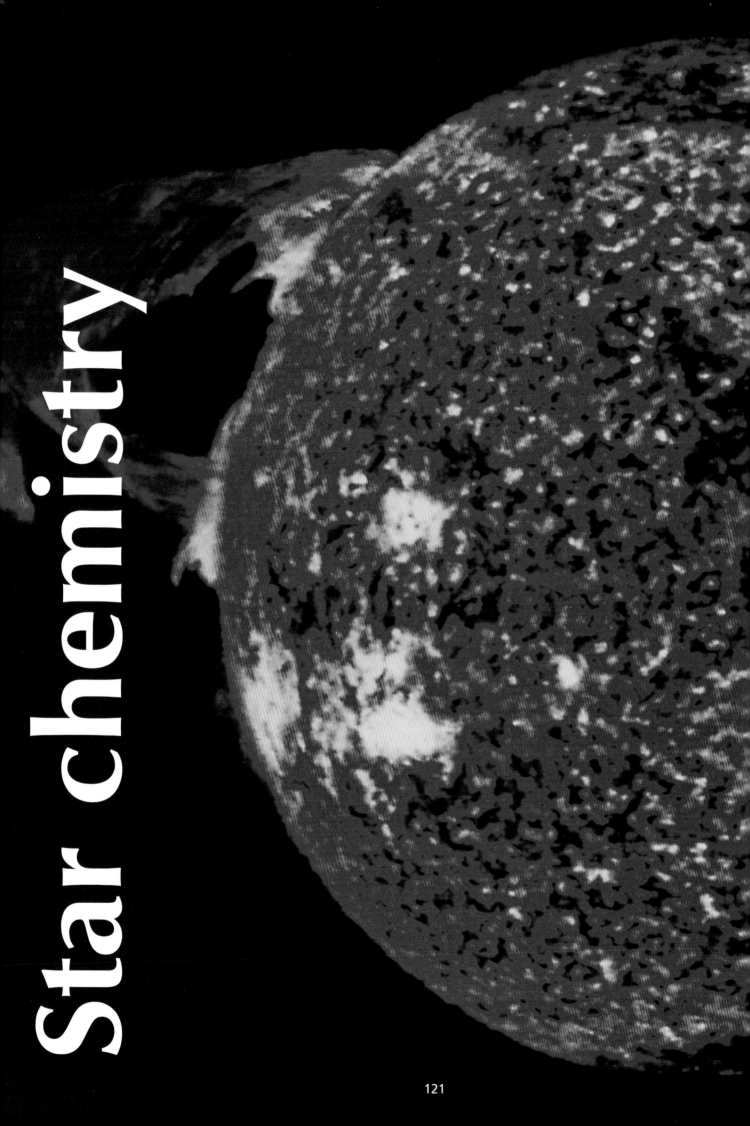

# Star chemistry

# RS•C

# Star chemistry

**Summary**

This material introduces astrochemistry. a relatively new area of science exploring the reactions among atoms and molecules in space.

| Resource name | Index | Type | Age range | Topic | Media |
|---|---|---|---|---|---|
| *Seeing space* | *08.01* | Paper based exercise – picture observation and questions | 11–16 | Spectral colours | |
| *Did you know? About hydrogen* | *08.02* | Information | 11–16 | Hydrogen | |
| *Key words* | *08.03* | Glossary | 11–16 | Glossary | |
| *The Interstellar Medium (ISM)* | *08.04* | Paper based exercise – picture observation and questions | 14–16 | Atoms and molecules | |
| *Did you know? About the Interstellar Medium* | *08.05* | Literacy / DART | 14–16 | What is in space? | |
| *Chemical reactions in the Interstellar Medium* | *08.06* | Literacy and data analysis questions. Interactive activity: Molecules in space | 11–14 | Interstellar medium | |
| *Debate! Can we really do 'space chemistry' on Earth?* | *08.07* | Literacy / DART and questions | 14–16 | The nature of science | |
| *The basis for life: analysing large molecules in space* | *08.08* | Molecular modelling and report writing | 14–16 | Molecular modelling | |
| *Did you know? What's in a name* | *08.09* | Information | 14–16 | Nomenclature | |
| *Did you know? About radio telescopes* | *08.10* | Information | 14–16 | Radio telescopes | |
| *The discovery of vinyl alcohol* | *08.11* | DART based on a press release | 14–16 post 16 | Vinyl alcohol | |
| *Person Profile: Dr Serena Viti* | *08.12* | Careers information | 11–16 | Careers | |

Key: ⬤ Interactive student activity  ⬤ Photocopiable and printable worksheet  ⬤ Projectable picture resource

Issue

Space can be explored by ground based telescopes. Experiments can imitate space conditions. What can these tell us about the origins of matter and the possibility of life elsewhere in the universe?

Chemical topics
- Introduction to spectral colours
- Covalent bond formation
- Making hydrogen molecules
- Mechanisms for making molecules
- Naming molecules
- Steady state systems.

Scientific enquiry issues
- How scientific ideas are presented by scientists
- Using evidence to answer a research question
- Consideration of the limitations of science.

Notes on using the resource

The activities are all free-standing, in that there are no prerequisites – each could be used independently of the others. The only exception is that **Debate! Can we really do 'space chemistry' on Earth?** as set makes more sense following the research paper activity.

The first two paper based activities could be used in a variety of settings where chemical elements and molecules are being introduced. Students can see that these exist in space by the spectral colours and other electromagnetic wavelengths emitted. Discussion could continue to show how the chemical elements form in stars.

**The Interstellar Medium (ISM)** introduces the idea of a steady state situation – hydrogen molecules are made and destroyed, but the amount of hydrogen molecules in space remains constant. Studying pictures and a research paper illustrate how scientists are trying to solve how molecules form in space conditions. This introduces the notion of bond formation, so could be used in a topic on chemical bonding, or perhaps as an introduction to chemical equilibrium. Students can debate the issue of whether Earth-based experiments can tell us more about space-based chemistry.

In **The basis for life: analysing large molecules in space**, students take on roles in a team of scientists researching the question 'Molecules in the ISM: Are these clues to life in space?'. Their findings are to be written up as a 700 word article for publication. The activity introduces molecular modelling, the Periodic Table and the formation of polymers as molecules on which life is based. The activity would provide good background and revision of chemical bonding, showing that prevailing conditions can influence bond type. The second part uses a press release describing the discovery of vinyl alcohol in space, pointing to the possibility that polymers may be able to exist.

## Seeing space

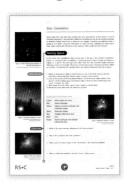

*Index 08.01*

Learning objectives
- Colours in pictures of stars and nebulae relate to chemical elements
- Hydrogen is the most common element in the universe
- Hydrogen atoms are the starting material for other chemical elements.

Time required
About 30 minutes

This activity uses spectacular optical telescope pictures of parts of the familiar constellation Orion. These can be presented as OHTs or used with a data projector for class discussion, in which case the activity may require less time as questions could be answered collectively. These pictures could be supplemented with others from the websites listed in resources, subject to copyright permission. Discussion could continue to explore the formation of the chemical elements from hydrogen atoms. Note that hydrogen atoms give the familiar red colour. Hydrogen molecules do not emit light in the visible region – see **The hydrogen problem**.

A completed observation table may include:

| Picture name | Colours | What makes the colours? |
| --- | --- | --- |
| The Orion Nebula | Red | Atomic hydrogen |
|  | Green | Molecular oxygen |
|  | Yellow | Hydrogen and oxygen mixed |
|  | Black | Clouds of dust, ISM |
| The Horsehead Nebula | Black | Clouds of dust, ISM |
|  | Red | Atomic hydrogen |
|  | Blue | Reflected starlight |
| Reflection nebula in Orion's sword | Blue | Reflected starlight |
|  | Red | Atomic hydrogen |
|  | Black | Clouds of dust, ISM |

Answers

1. Atomic hydrogen.

2. The ISM surrounds all stars and planets, so can be found everywhere. Atomic hydrogen is the simplest substance and the starting point for making other chemical elements, so is very common in the universe.

3. The black shape is caused by a large cloud of dust.

4. The light from these objects is too faint to be detected by the naked eye.

The picture of the Orion Nebula was taken using 3–5 minute exposures of a photographic film attached to the Anglo-Australian telescope.

## *Did you know? About hydrogen*

*Index 08.02*

## *Key words*

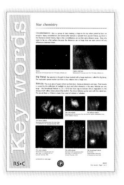 

*Index 08.03*

## *The Interstellar Medium (ISM)*

  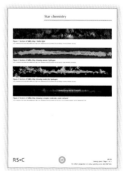

*Index 08.04*

Learning objectives
- A 'steady state' situation means the rate of formation = rate of breakdown
- The Interstellar Medium (ISM) is rich in molecules and chemical reactions
- Pictures of space in non-visible regions give extra information about chemical reactions
- Hydrogen molecules do not form by accident but by one or more mechanisms
- To explore how astrochemists carry out and publish research.

### The hydrogen problem

The 'hydrogen problem' is that there are more hydrogen molecules in space than can be explained by formation by collision of atoms alone. Molecules of hydrogen are formed and destroyed (by UV radiation) at approximately equal rates, so a steady state exists. Atoms of hydrogen are too far apart in space for enough effective collisions to occur at the rate needed to maintain the observed amount of molecules. This means that the molecules must form in a different way.

The Interstellar Medium (ISM) has relatively recently been found to be a rich source of molecules and chemical reactions. The ISM is the region between the stars, and contrary to popular belief, is not empty, but has dust and gaseous molecules in it. The density of the ISM can vary, so the terms dense and diffuse are used to describe the bulk differences. Scientists observe the ISM on a gross scale using telescopes and imitate microscale events to help to explain these observations.

### Seeing the Milky Way

Time required
About 30 minutes

In this activity, students observe pictures of the same part of the sky and realise that molecular hydrogen is formed in the ISM, not in stars.

Students need access to Figures 1, 2, 3 and 4. These are of the same part of the Milky Way. Students based in an area relatively free of light pollution will be able to see the Milky Way by standing outside for a few minutes and looking for the bright band of stars across the sky.

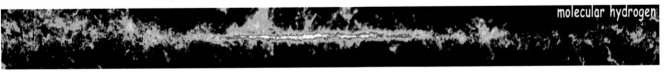

*Figure 1 Section of Milky Way: Visible light*

Source: Astrophysics Data Facility, NASA Goddard Space Flight Centre. (Reproduced with kind permission from Axel Mellinger, University of Potsdam, Germany).

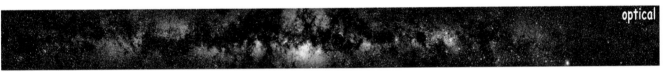

*Figure 2 Section of Milky Way showing atomic hydrogen*

Source: Astrophysics Data Facility, NASA Goddard Space Flight Centre. (Reproduced with kind permission from Dap Hartmann, Leiden observatory, Netherlands).

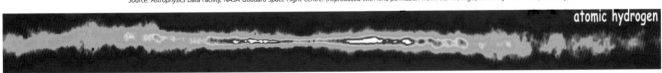

*Figure 3 Section of Milky Way showing molecular hydrogen*

Source: Astrophysics Data Facility, NASA Goddard Space Flight Centre. (Reproduced with kind permission from Axel Mellinger, University of Potsdam, Germany).

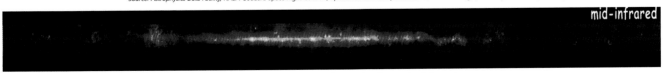

*Figure 4 Section of Milky Way showing complex molecules under infrared*

Source: Astrophysics Data Facility, NASA Goddard Space Flight Centre. (Reproduced with kind permission from Steve Price, Air Force Research Laboratory, Hanscom, Massachusetts, USA).

The pictures could be presented as OHTs to help overlay areas. Questions could be answered by class discussion. The aim is to see that the region in which molecular hydrogen is found (Figure 3) overlaps with the areas where complex molecules are found (Figure 4). This suggests chemical reactions are going on in the Interstellar Medium, the dark areas of dust between the stars. The point of the next activity is to think about what reactions these might be and how these happen. More information about the ISM is provided in **Did you know? About the Interstellar Medium.**

Answers

1. Atomic hydrogen is found in regions where there are stars.

3. Molecular hydrogen is found in the Interstellar Medium, in the blackest parts of the pictures – where there are no stars.

4. Complex molecules are also found in the Interstellar Medium. This tells us that chemistry is happening in the ISM.

## *Did you know? About the Interstellar Medium*

*Index 08.05*

Answers

A single covalent bond. The bond forms when the atomic orbitals of two hydrogen atoms overlap. The electron in each of the two orbitals becomes shared between the two atoms. This means that the region of space in which the two electrons could be found is now between two atoms. The electrons are described as shared.

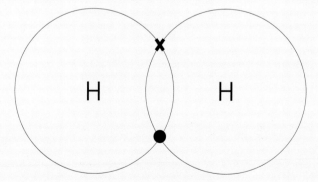

The chances of two hydrogen atoms meeting in space and forming a molecule are very small. To form a molecule, the atoms must be close together and not bounce apart due to repulsion between the two negatively charged particles (electrons).

3. The black parts of the sky are filled with dust.

## Chemical reactions in the Interstellar Medium

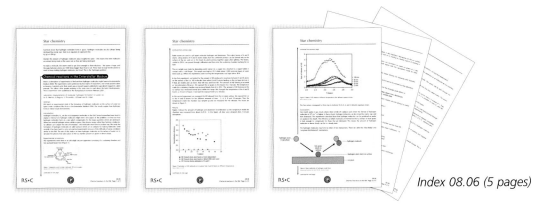

*Index 08.06 (5 pages)*

Learning objectives
- To understand how scientific research is carried out
- To understand how simple reactions occur.

Time required
About 60 minutes, perhaps longer for the debate.

In this activity students are led through a research paper seeking to explain how hydrogen molecules might form in the ISM.

The research paper presented is an adapted version of an original article. The reference for the original is given in the resources list. The text was amended with the agreement and help of one of the authors, Valerio Pirronello.

In the experiments, ice was made by building up layers of water vapour on a cold surface in a scattering chamber made from stainless steel. The cold surface was a copper disk attached to a copper holder in contact with a 'cold finger'. The temperature was kept at about 10K while this was happening. Beams of hydrogen and deuterium atoms were made using radiowaves to break bonds in the hydrogen and deuterium molecules. The pressure inside the chamber was reduced to an ultra high vacuum. A diagram of the apparatus is shown in Figure 1.

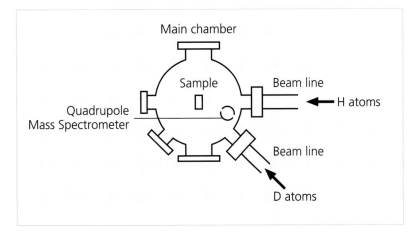

*Figure 1 Apparatus used to make molecular HD on ice grains*
Reproduced with permission from Valerio Pirronello, University of Catania.

The researchers carried out several experiments with this set up. Two are reported here. In the first, the rate at which hydrogen and deuterium atoms combined on the surface of the ice was measured. This is reported in Figure 2. The x-axis shows the temperature inside the chamber. This was increased slowly. The radio signal given off by HD molecules was detected and the values shown were obtained. These are given as recombination efficiency on the y-axis. The results shown in the unfilled circles are those obtained using atoms at 200K and the filled circles are those obtained using hotter beams at room temperature. The triangles show the levels of HD detected while the atoms were being adsorbed. This experiment mimics the formation of hydrogen molecules on ice grains in space. The researchers were trying to find out if the experimentally determined rate at which the molecules of HD form might match theoretical predictions.

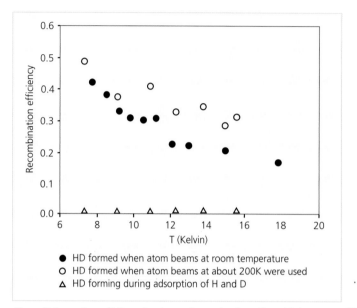

Figure 2 *Formation of HD molecules on ice grains from H and D beams at different temperatures.*
Based on Manico et al, 2001.

In the second experiment, the rate of release of the HD was measured after the ice was exposed to the beams for different amounts of time. The temperature in the chamber was increased to higher values and the amount of HD in the chamber was measured. The results (see Figure 3) show that more HD was produced after longer HD exposure times, and that the most HD is released at about 25K. This experiment shows that hydrogen molecules are released from the ice grains and into the ISM.

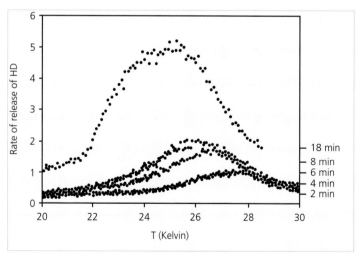

Figure 3 *Rates of HD release at different temperatures after different exposure times.*
Based on Manico et al, 2001.

This may appear to be a difficult activity, probably best attempted with relatively able and motivated students. The activity is designed to help students understand the process of scientific research. There is also the opportunity to learn about new techniques and the role of experimental modelling in developing scientific theories.

Teachers could impress on the students the novelty of the experiment – here, scientists are trying to mimic ISM conditions, using very cold temperatures close to absolute zero and low pressures. This is very difficult and costly to create on Earth! Also, the collaborative work is important – experiments often involve more than one scientist, as contributors take on different roles. For example, here, some may have worked more on the experimental techniques, processing the data and analysis of the data. The original paper and others in this field necessarily include complex calculations, which, of course, are omitted. This leads to presentation of quite a truncated view, but the point of the work should be clear from the account given.

Answers

1. See marked up text.

2. Abstracts provide a summary of the research. Scientists can read the summary to decide if the report is useful for their own work without spending time reading the whole document.

3. Hydrogen molecules are the most important molecules in the ISM. They want to find out how the molecules form at a rate equal to the rate they are being destroyed. This is the 'hydrogen problem'. Also, so far only theoretical calculations had been done. The practical experiment tested the theory. Doing an experiment is helpful to try and find out what is really happening and to confirm theoretical predictions.

4. (a) ultra high vacuum and (c) cold temperatures were needed to help imitate space conditions. Also, the vacuum reduces the number of molecules around, so that only hydrogen and deuterium atoms from the beams would be in the chamber. The atomic beams of hydrogen and deuterium (b) were needed to provide the atoms. Two beams of hydrogen could not be used because otherwise the scientists would not be able to tell the difference between hydrogen molecules which already existed and new ones which formed as a result of the mechanism on the ice.

5. The most HD seems to be formed when the temperature is coldest, at 7K. The amount of HD formed seems to be higher with the beams at 200K. Colder atoms may stick more easily to the ice.

6. (a) The most HD was released after 18 minutes exposure and at about 25K.
   (b) More HD had a chance to form due to the longer exposure time. Atoms have a better chance of meeting when more are stuck on the surface.

7. The scientists found that HD formed on the ice grain surfaces and that this was released. The results support their conclusions.

8. Scientists work in teams because they each have different skills and knowledge. They also contribute different ideas. Experiments like this can be very expensive so collaborations between scientists in different countries can reduce the costs. The technician would have helped to maintain the equipment and help to keep it working.

9. The answer to this will depend on what your own practice is. Often, investigations make a prediction, there is a description of the experiment together with a diagram, results are presented and conclusions are drawn. Comparisons are likely to be quite close.

## Debate! Can we really do 'space chemistry' on Earth?

*Index 08.07*

### Learning objectives
• To discuss whether trying to replicate on Earth conditions found in space is valid.

### Time required
30 minutes

This activity uses the experiment in the resource **Chemical reactions in the Interstellar Medium**. Students will need to have completed this activity prior to attempting the debate exercise.

## The basis for life: analysing large molecules in space

*Index 08.08*

### Learning objectives
• A large number of molecules have been found in space
• Revision of covalent bonding and formation of polymers
• Some molecules found in space are found on Earth
• Some space molecules could not exist on Earth
• Bond types depend on the local conditions
• Writing a scientific article based on experimental results.

### Time required
60 minutes to work in roles preparing material. Extra time to prepare the article.

Students should work preferably in teams of three or four. Three is preferable – in this case ask students to take the tasks of element analysis, molecular modelling and looking at practical techniques, working together on the conclusions.

Use the **Molecules in space table** as a basis for revising ideas about chemical bonding and the formation of polymers. The table shows 60 of the molecules which have been detected in space, including some with unsaturated (double and triple covalent) bonds. Two key points can be drawn out of this:

The unsaturated molecules could form the basis for making polymers, as under the right conditions, the unsaturated bonds could break, allowing monomers to bond together. However, the chances are that in space there would be insufficient monomers in one locality and polymer molecules may be broken up by UV radiation anyway.

Compounds may have different bond types if they are made in dramatically different conditions such as outer space. Potassium chloride, for example, could be covalent in space, because of the local conditions and also because there are insufficient particles present to form an ionic lattice. Try to show students that bond types depend on local conditions, *eg* energy availability and presence of other particles. This enables students to move away from a stereotypical view of bonds as either covalent or ionic, but to consider a wider view.

Each activity will require students to do some more fact-finding. Access to the internet or reliable texts will be required. The resource list at the end of the unit will help provide starting points. A standard molecular model kit *eg* Molymod, should be adequate for making the molecular models here. Making triple bonds can be difficult unless the long grey links are used.

Answers

### Element analysis

1. Hydrogen, carbon, oxygen, nitrogen are basic elements for supporting life. Yes, these are present in space molecules.

2. The Noble Gases (group O) do not appear. These elements have atoms which do not readily form compounds.

3. The total number of molecules in the table is 60. Students may find this surprising because they did not realise that there would be so many. Others have been detected – the total number is around 125.

### Molecular modelling

1. The presence of unsaturated bonds means that molecules could join up with each other making long molecules.

2. Students to work on this themselves.

3. Atom – neutral particle in which the numbers of protons and electrons are equal.
   Ion – charged particle in which the number of protons may be greater or less than the number of electrons.
   Molecule – particle made when two or more atoms bond together with one or more covalent bonds.
   Radical – a particle with an electron available which has not been used to make a bond. An atom can be a radical.

### Practical techniques

1. Radio telescopes are used to detect molecules in space so far away.

2. The signals can be compared to those given out by the same molecules present on Earth.

3. Large molecules like those present in living things on Earth would need to be detected.

### Making conclusions

1. There are more smaller molecules because the chances of these forming in space are greater. It is much more likely that two atoms could meet than three or more. Also, larger molecules are more likely to be broken up by UV radiation.

2. The monomers would need to collect together in one place and the conditions would need to be right to make them react together.

3. Possibly – some larger molecules are needed for life, but not all large molecules indicate life.

4. Scientists are looking for regular signals from space which would indicate the presence of life. Space ships could also be sent out to look. This is the 'sci-fi' question. The issue of scientists maintaining an open mind on this issue could be discussed.

## *Did you know? What's in a name*

*Index 08.09*

## *Did you know? About radio telescopes*

*Index 08.10*

### *The discovery of vinyl alcohol*

*Index 08.11*

Learning objectives
- Extracting information from a press release
- Application of chemical bonding and isomerism to a new situation.

Time required
About 30 minutes

Answers

1. An accurate name from the **What's in a name** information, would be etheneol. Looking at the **Molecules in space** table shows 'hydroxyethene'.

2. An isomer is a molecule with the same number of atoms of each type as another, but in a different arrangement. Isomers of vinyl alcohol are ethanal and ethylene oxide. Other families/ pairs of isomers are ethanoic acid/methyl methanoate (8 atoms) and ethanol/methoxymethane (9 atoms).

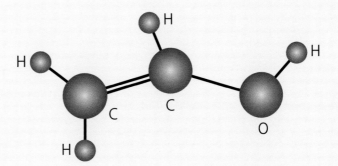

*(Reproduced with permission from National Radio Astronomy Observatory/AUI, Virginia, USA.*

3. The sections are in paragraph 4 – '…molecules lock horns…' and in paragraph 5 beginning with '…molecules could form on tiny dust grains'. Scientists are not satisfied with the dust grains mechanism because this would not give free molecules – they would stay stuck to the ice grains.

4. The molecule is a link towards understanding how complex organic compounds may form (paragraph 2). The double bond in the molecule suggests it could polymerise under the right conditions.

5. Astrochemistry studies chemical reactions in space, suggesting mechanisms for how molecules could form. Astrochemists detect molecules such as vinyl alcohol and then work on suggestions for making these, such as those described in paragraphs 4 and 5.

6. Another mechanism is suggested in the second to last paragraph; 'The processes evaporating ...' The uncertainty is shown at the end of the third to last paragraph; 'Turner warns...' and in the last paragraph, 'Scientists plan more work...'.

Science is about seeking answers to problems thrown up by data, and that hard and fast answers are not always apparent. Science facts are won by observation and experiment, but can change as new information arises.

## *Person profile: Dr Serena Viti*

*Index 08.12*

## *Further information, references and source material*

These websites have information which is relevant to this topic:

*www.aao.gov.au/images.html/* (accessed November 2003) gives more pictures of stars taken from the Anglo-Australian Telescope by David Malin.

*www.lifeinuniverse.org/noflash/Molecules-02-02.html* (accessed November 2003) gives information about the molecules in space.

*www.seti.org* (accessed November 2003) is the website for the Search for ExtraTerrestrial Intelligence. This has interesting information about the origin of life in the universe.

*www.beagle2.com* (accessed November 2003) is the website relating to the Beagle expedition led by UK scientists to land on Mars in 2003.

*www.nrao.edu/education* (accessed November 2003) is the National Radio Astronomy Observatory website which gives an introduction to radio telescopes.

*www.rog.nmm.ac.uk/* (accessed November 2003) is the home page of the Royal Observatory Greenwich, which maintains an update of the latest astronomy news.

*www.pparc.ac.uk* (accessed November 2003) is the UK's Particle Physics and Astronomy Research Council website and is full of useful information including pictures.

*www.star.ucl.ac.uk/* is the home page of the University College Astronomy Group, where Serena Viti works.

**Other resources**

Various molecular model kits can be purchased from educational equipment suppliers. The Molymod kits are available by ordering on the web at *www.molymod.com/* (accessed November 2003) and
Spiring Enterprises Limited
Gillmans Industrial Estate
Natts Lane
Billingshurst
West Sussex
RH14 9EZ
Telephone +0044 (0) 1403 78 23 87
Fax: +0044 (0) 1403 78 52 15
Phone lines are open at the following times: Mon - Fri - 10am - 5pm (UK time)
*email@molymod.com*

Other companies are:
Cochranes of Oxford Ltd
Fairspear House, Leafield, Oxford OX8 5NY
+44 (0)1993 878 641
*www.cochranes.co.uk*

*www.indigo.com/models/molymod-molecular-model-kits.html* (accessed November 2003)
Student sets 008 or 009 would be adequate for this unit. These include links which enable triple bonds to be made.

**References**

G. Manico, G. Raguni, V. Pirronello, J.E. Roser, and G. Vidali, *Laboratory measurements of molecular hydrogen formation on amorphous water ice, The Astrophysical Journal*, 2001, 548, 253–256.

S. Viti, *What a dusty universe! Chemistry Review*, December 2000, 30–33.

The Economist, *A survey of the universe*, 5th January, 2002, 51–62.

T. Ferris (Editor), *The World Treasury of Physics, Astronomy and Mathematics* London: Little Brown and Co, 1991.

J. Gribbin, *Companion to the Cosmos*, London: Little Brown and Co.

# Cleaning chemistry

LUX
SHOWER

**NEW**

*Shimmering sea*

**SHOWER GEL**
DOUCHEGEL

sea minerals

Zeemineralen

**Summary**

Soaps and detergents are used everyday for a range of different cleaning purposes. What these are, how they are made and the problems associated with their use are explored in this collection of resources.

| Resource name | Index | Type | Age range | Topic | Media |
|---|---|---|---|---|---|
| *Testing shower gels and soaps* | 09.01 | Class experiment and questions | 11–14 | Testing shower gels and soaps |  |
| *Did you know? About shower gels and soaps* | 09.02 | Information | 11–16 | Shower gels and soaps | |
| *How do soaps and detergents work?* | 09.03 | Demonstration and questions Interactive exercise | 11–14 | Surface tension | |
| *Key words* | 09.04 | Glossary | 11–16 | Glossary | |
| *Extension activity* | 09.05 | Investigation | 11–14 | Testing properties of soaps and detergents | |
| *Look at the label!* | 09.06 | Class paper based data collection activity / survey | 11–14 | Product analysis | |
| *Cosmetic ingredients database* | 09.07 | Database | 11–16 | Database | |
| *Product analysis sheet* | 09.08 | Data collection | 11–16 | Data collection | |
| *Making soap* | 09.09 | Class experiment and questions | 11–16 | Making soap | |
| *Did you know? About detergents* | 09.10 | Literacy / DART and questions | 14–16 | Detergents | |
| *The history of soap* | 09.11 | Literacy / DART and questions | 11–16 | History of soap | |
| *Using soap and soapless detergents* | 09.12 | Discussion exercise | 11–14 | Soap and soapless detergents | |
| *Forever blowing bubbles* | 09.13 | Planning an investigation | 11–14 | Bubble investigation | |

Key: ⬤ Interactive student activity    ⬤ Photocopiable and printable worksheet    ⬤ Projectable picture resource

Issue

Soaps and detergents are used daily and their properties taken for granted. What are they made from and should we use as much?

Chemical topics
- Measuring pH
- Action of surfactants on grease
- Reaction between acid and alkali to make soap
- Disposing of detergents and soaps
- Making mixtures.

Scientific enquiry issues
- Planning a scientific investigation
- Collecting survey data
- Discussion of environmental problems.

Notes on using the resource

The activities are all free standing and could be used to support existing materials on topics such as chemical reactions, chemistry in the environment and acids and alkalis.

Teachers might like to note the common misuse of the terms soap and detergent; all soaps are detergents but not all detergents are soaps, *ie* there are soapless detergents. The phrase used in this resource will predominantly be soap and soapless detergents, however where the word detergent is used alone this is taken to mean both soap and soapless detergents.

**Testing shower gels and soaps** provides an introduction to soap and soapless detergents. Teachers could ask students to bring in soap and shower gel samples from home to test pH values and to disperse on an oil film. **How do detergents work?** is a demonstration to show how soap and soapless detergents act to break down surface tension.

In **Look at the label!** students survey the ingredients of at least three different soap or soapless detergent products. The **Cosmetic ingredients database** is needed to help with this. The activity introduces terms relating to making mixtures, as well as getting students to relate the ingredients to the claims made about the products.

**Making soap** is an experiment for the students to make their own soap using a simple procedure. This involves reacting a fat with sodium hydroxide solution. The procedure is quite lengthy, but satisfying as the results can be good.

**The history of soap** is a short text-based task introducing the history of soap followed by questions. This activity could be done in conjunction with **Making soap**.

Finally, in **Using soap and soapless detergents** students are invited to consider the environmental implications of using soaps so regularly. This is done by discussion - six different views are presented and students are asked to consider these and rank them in order of agreement, but to agree the rank order within a group first.

## Testing shower gels and soaps

*Index 09.01*

---

**Learning objectives**
- Understanding and measuring pH
- The action of surfactants on grease

---

Time required

About 50 minutes. 60 minutes if the demonstration **How do soaps and detergents work?** is included.

Students could bring a range of detergent samples from home *eg* bars of soap, shower gels etc. These can be solid soap bars or liquids. Students may be invited to prepare a small sample for the rest of the class to use. They can then keep the 'stock' bottle for reference and to avoid this being contaminated or used up. This will give a broad range of substances to test. Alternatively, a smaller range could be bought specifically for this experiment. The ingredients lists are also required, which are found on the bottles of liquids, but on the wrappers of solid soap bars, so care should be taken to bring wrapped soap bars. Results for a range of products are given below, together with answers to the questions.

Apparatus and equipment (per student or group)
- About 200 cm$^3$ distilled water
- 5-6 test-tubes
- Test-tube rack
- Universal Indicator solution and/or pH paper
- Dropping pipettes - one for each liquid tested
- Potato peeler to take shavings from solid soaps
- Petri dish
- Ruler
- About 50 cm$^3$ vegetable oil
- Eye protection
- Copy of results table
- Piece of dark coloured paper to sit underneath the petri dish (10 cm x 10 cm square)
- Shower gels and soaps to test.

This assumes a maximum of 6 tests per group.

Notes on the experiment

- If test-tubes are to be used repeatedly, they should be rinsed thoroughly.
- Most soaps will have pH values between 6–8. Note any that are marketed as 'low pH' such as Johnson's pH 5.5 – ask students what the benefits might be, or at least what is claimed for this product.

- Encourage students to carry out the oil test in exactly the same way each time. The dark paper underneath makes the ring of soap substance more visible.
- The pH test may require access to both Universal Indicator solution and pH paper. In practice, coloured substances are diluted by the addition of water, so work well with the Universal Indicator solution, but care needs to be taken to use small quantities. Obtaining accurate pH values with paper is more difficult than with solution.

Answers

1. Answers will vary depending on samples tested. Most soaps have pH values between 6-8.

2. Answers will depend on the ingredients contained in the test samples. The middle column of the Cosmetic ingredients database will help with this. Look for 'sodium hydroxide' (named 'sodium isethionate' on labels); citric acid; ammonium compounds (in shampoos). Encourage students to link the ingredients with the observed results.

3. Answers will depend on the samples tested. The surfactants in the soap solutions make it spread. Those samples containing the highest concentration of surfactant are likely to make it spread furthest. Students should relate the results to ingredients which are surface active or surfactants – these can be identified from the middle column of the database.

4 and 5. Answers will depend on the samples tested - the suggestion might be that the substance which produces the largest diameter spread in the oil test will be the best cleaner.

6. Any suitable suggestion with controlled variables *eg* wash standard sized equally dirty materials under the same conditions.

Sample results

| Product | pH | Oil test / cm |
|---|---|---|
| Dove body wash | 6.0 | 4.0 |
| Simple shower gel | 6.0 | 8.0 |
| Johnson's pH 5.5 shower gel | 6.0 | 8.0 |
| Tesco Value shower gel | 6.0 | 4.4 |
| Tesco Shower gel | 6.0 | 5.5 |
| Oil of Olay body wash | 6.0 | 4.0 |
| Salon selectives shampoo | 6.5 | 3.5 |
| Trevor Sorbie shampoo | 7.5 | 3.3 |
| Garnier Healthy hair shampoo | 7.5 | 2.0 |
| Tesco Value soap | 7.5 | 0.5 |
| Pears transparent soap | 8.5 | 0.5 |
| Dove Cream Bar soap | 7.0 | 0.5 |
| Self-made soap (using lard) | 9.0 | 0.5 |
| Imperial leather soap | 7.5 | 2.0 |
| Tea tree oil soap | 6.5 | 0.5 |

## Did you know? About shower gels and soaps

*Index 09.02*

## How do soaps and detergents work?

*Index 09.03*

### Learning objectives
- The role of soaps and detergents in reducing surface tension
- How detergents interact with both grease and water.

### Demonstration: The incredible floating drawing pin

#### Time required
- About 20 minutes

The demonstration will help students understand how soaps and soapless detergents work. Sleight of hand is required, so practice is strongly recommended.

#### Apparatus and equipment
- Plastic trough, *eg* margarine tub filled with water
- Drawing pin or paper clip
- Cocktail stick pointed at both ends
- Small amount of liquid soap or detergent.

#### What you do

- Dip one end of the cocktail stick in the detergent. Do not let anyone see you do this.
- Float the drawing pin on the surface of the water. This takes practice. Get everyone to watch carefully what you do next. Very gently push the non-soapy end of the cocktail stick into the surface of the water so that the pin stays on the surface. Do this repeatedly so that it looks like this happens every time.
- Discuss surface tension holding the drawing pin in place (NOT forces from the water). Teachers could say that care is needed not to disturb the surface tension, or the pin will sink.

- Now challenge a student to repeat this – explaining that you yourself are so skilled that no-one else could do this without causing the drawing pin/paper clip to sink. When you get a volunteer, reverse the cocktail stick so the student has to use the detergent-soaked end to prod the surface of the water. After 2–4 attempts, the drawing pin will sink.
- Repeat this with another student, or immediately confess.

Explanation

The pin is not floating but sitting on the surface tension of the water. At the surface the forces of attraction between the molecules are unbalanced. In the liquid they are balanced. This is why a beaker/cup can be filled above the brim. The soap on the stick lowers the surface tension because soap molecules get between the water molecules reducing the forces of attraction at the point where the stick hits the water. The surface tension is then insufficient to support the mass of the pin/clip, which then sinks.

The demonstration shows that soap/detergent lowers the surface tension of water. This is the basic principle on which they work.

Answers

2.

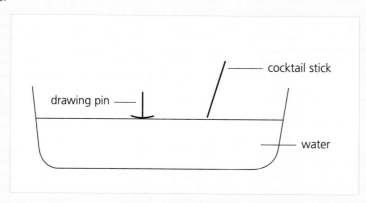

3. The detergent reduced the surface tension so the drawing pins could not be supported.

**Demonstration: Mix 'n' match**

This is an effective way of demonstrating how a soap or soapless detergent works.

Apparatus and equipment
- 2 x 1 dm³ beakers
- 600 cm³ water
- 300 cm³ vegetable oil
- Few drops of detergent substance
- 2 spatulas or splints to stir.

What you do

- Pour equal amounts of water into the two beakers. Pour oil into both - the amounts do not need to be precise, but enough to make a visible layer and such that both beakers appear to have equal quantities.

143

- Add a few drops of a detergent substance to one of the beakers.
- Stir the liquids in both beakers vigorously. This could be done by volunteers. The layers of oil and water will mix. The mixture including the detergent will form small micelles of oil and will look much more 'mixed' than the oil/water with detergent.

Answers

2. The soap/soapless detergent molecules react with both the oil and water molecules, enabling them to mix together. Besides being surfactants, the detergents in the product act as emulsifiers. This means making two liquids mix together which normally make layers. The mixture is an emulsion.

   The description emulsion can also be applied to liquid soap substances. These often include one or more emulsifiers to ensure that added oils (usually to 'soften' or moisturise skin) mix with other ingredients.

3. The detergent molecules have one part which can bond with water molecules while the other end can bond with oil and grease. This means that the detergent molecules can lift grease from a surface while the other end of the molecules bond with water. The water dissolves the detergent /grease complex and takes it off the skin/fabric/dish.

## *Key words*

*Index 09.04*

## *Extension activity*

*Index 09.05*

## Look at the label!

*Index 09.06*

Learning objectives
• To recognise the wide variety of chemical substances in household products.

Time required
About 40 minutes

The time required to research the ingredients in three products is rather elastic - in practice, students should be able to manage more than three. Students will require unrestricted access to the **Cosmetic ingredients database** for this activity. Extra copies of the analysis sheet will be needed if students are to work on more than three products. The products could be the same as those tested in the first experiment, but this activity could be run in isolation.

It is best if the products include one or more used by the students regularly. This will give the most impact. Students will need the ingredients lists and the costs of the products. The costs are often given as 'price per 100 ml' or 'per 100 g' on supermarket shelves. These units have been used on the analysis sheet.

Notes on using the Cosmetic ingredients database

• The database was prepared using the ingredients lists from the range of products in one supermarket, including the supermarket's own brand. The list may not be fully comprehensive for the products available in your area. Missing ingredients can be traced by using the resource materials (refer to References list).

• 'Aqua' is the first ingredient in all liquid products and appears quite high up in the list on many soap bars because this is the basic solvent on which products are based. Aqua is the manufacturers' name for water. The task is more interesting if water or aqua is excluded.

• Sodium tallowate or sodium palmitate will be the main ingredient in a solid soap bar.

• All products will contain one or more surfactants. Liquid products will also include emulsifiers. Solid soap bars may also include skin softeners and oils.

• Ingredients are listed in order of mass from highest to lowest. Manufacturers have to include all ingredients, but are not obliged to state the quantities. Thus, it is possible for an expensive product to include only a tiny amount of a chemical which allows the label to state, for example, 'includes herbal extracts', even if the amount is too small to have a positive effect.

• Some ingredients are the same in all or most products because these create the cleansing properties needed. Ingredients differ depending on the claims made for the product, so a 'low pH'

145

shower gel should contain an acidic compound which will lower the pH. A product claiming to be 'kind to sensitive skins' may include skin softeners or natural products.

- Some chemical names are complicated and have similar spellings to others. Students will need time to write these down and should be encouraged to check for accuracy.

- Some names are very similar. An example is 'cocoamide DEA', which is also known as 'cocoamide MEA'. Where the properties are basically identical, these are given as one entry.

- Cross-references are made to the main groups of ingredients, namely emulsifiers, cleansers, surfactants, acids, alkalis, etc. These names are placed in the central column. Natural ingredients may not have identifiable characteristics in which case the centre column is left blank.

- The third column may be blank if no other information was available.

- The information provided is necessarily selective. If further details about an ingredient are required, please consult the references below or search the web for individual substances.

- Cosmetics companies do not use systematic names for compounds. This is one reason why the names can be so difficult. Also, some chemicals are given names which hide their true identity – a good example is sodium hydroxide, which may appear as 'sodium isethionate'. Ask students why this is done – one answer is that although manufacturers are forced by legislation to list the ingredients, they actually do not want to tell us what is really there!

## Answers

1. The ingredients are needed to give the product the expected properties. For example, the surfactant is needed to reduce the surface tension – this has the cleaning properties. An emulsifier keeps oils and water-based substances mixed; the colour makes the product look nice; other ingredients are there to soften the skin, retain moisture or maintain pH. Antiseptics and anti-bacterial agents prevent the product from growing microbes while on the shelf.

2. This depends on the products in the students' lists. Probably all the liquids will have two or more of these characteristics. The solid soaps could be described as solid emulsions. Discuss the differences and similarities between these terms with students. See **Key words**.

3. This depends on the products in the students' lists. Students will be able to identify ingredients added, for example, as skin softeners and link these to 'kind to skin'; those which are 'refreshing' contain astringents such as witch hazel; 'antibacterial' cleansers such as anti-acne products contain tricloban/triclosan.

4. Cheaper supermarket-branded products tended to include some quite unpleasant chemicals such as formaldehyde (methanal) and a wide range of perfumes and colouring pigments. More expensive products used milder surfactants, natural oils and also included products tested as non-irritating.

5. Students with sensitive skins may now be able to work out which products irritate them and why. They would be advised to find products which use 'sodium laureth sulfate' rather than 'lauryl' for example.

6. The answer almost certainly is no. Society places an expectation on us to use these every day, but we probably use more than we really need. This could be a good discussion.

7. They did not really bother to keep very clean. Washing was a bit of a luxury and was probably done in rivers and streams, then more formally in bath-houses. People used heavily scented perfumes to mask body odours. Also people became used to smells. Deodorants were not invented until the 1950s and showering daily has developed as a fashion in developed countries over the last 20 years. Students may have grandparents who recall the 'weekly bath' by the fire.

## *Cosmetic ingredients database*

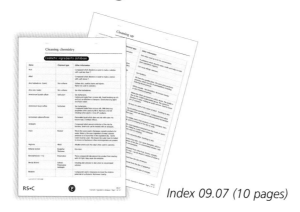

*Index 09.07 (10 pages)*

## *Product analysis sheet*

*Index 09.08*

## *Making soap*

*Index 09.09*

Learning objectives
• To produce a soap by the reaction of a fat with sodium hydroxide.

# Cleaning chemistry

Time required
Minimum 60 minutes, less if the apparatus is already prepared

Apparatus and equipment (per student or group)
- Two thermometers 0–100 °C
- 250 cm$^3$ beaker
- Two rubber spatulas
- Four plastic moulds, *eg* empty small margarine tubs or small plastic bowls
- Rubber/nitrile gloves
- Eye protection – goggles
- Balance
- Heatproof container, *eg* 1 dm$^3$ beaker
- Two other containers for weighing the water and sodium hydroxide
- Grease for the moulds, *eg* small amount of the fat, oil spray or petroleum jelly
- Apron or labcoat
- Bunsen burner
- Heatproof mat
- Tripod and gauze
- 454 g dripping OR lard OR solid vegetable fat
- 57 g sodium hydroxide (**Corrosive**)
- 142 g water
- 5 g essential oil (optional).

Pre-lesson preparation and notes on the experiment

The fat could be bought from a supermarket. This is cheap and readily available. Lard and dripping are most effective. Trex produces very soft soap.

An essential oil can be bought in many chemists, health food shops and branches of The Body Shop. A strong flowery fragrance such as geranium works well. These are expensive, but only small amounts are needed. This is an optional extra.

The method provides the instructions for using one complete block of fat. Although this seems a large quantity, the soap made is relatively compact and small, so it may be best to retain this size. You could share a block between two reaction mixtures, but the corresponding masses of other ingredients must be halved also.

Students could bring their own fats to the lesson and make their own soap. However, the time required to reach 'trace' (see below) is long – you may prefer to set up the reaction mixture and have the mixture timed to reach 'trace' during the lesson. Students can start off their own mixtures in the lesson itself, watch trace occurring in the pre-prepared mixture and pour their own mixtures into moulds for the reaction to be completed unobserved.

Teachers may be uneasy about allowing students to handle sodium hydroxide and the solution. If so, one option is to make the sodium hydroxide solution in advance, using 57 g NaOH to 142 g water (this is a 10 mol dm$^{-3}$ concentration). Alternatively, set the experiment up as a demonstration alone, using a pre-prepared solution. Whatever method is followed, eye protection (goggles not safety spectacles) is essential, as is careful supervision.

It is best not to use an aluminium-based container.

It is not recommended that students take their soap samples home. After the appropriate curing period, students could imitate a handwash keeping rubber gloves on.

Students could test the pH of the soap made in the method described in **Testing shower gels and**

**soaps**. Immediately after solidification the pH will be about 8-9, or even higher. As the saponification reaction continues slowly the pH will fall, as the sodium hydroxide content will be reduced. After several weeks the pH should be in the safe range, which is 6-8.

There are many recipes for making soap. Several references are given at the end for this if you wish to extend the soap-making activities beyond that described here.

### Safety

Wear goggles – this is essential due to the use of solid sodium hydroxide.
Wear gloves – preferably nitrile because latex allergies are common.
This experiment should ony be attempted if teachers are confident that all students will adhere to all safety requirements and procedures.

### Answers

1. The reactants are the fats in the block of fat and the sodium hydroxide.

2. At trace the reaction between sodium hydroxide and the fat has taken place to the extent that a lot of soap molecules are produced. The quantity is sufficient to cause the mixture to go opaque.

3. The energy involved is relatively low. The molecules of fat are large and take time to mix with the sodium hydroxide particles.

4. Sodium hydroxide may not be listed because it is all used up in the reaction. Traces of sodium hydroxide may be found in the soap but this is not usually sufficient to cause problems. Also cosmetic companies may hide the presence of sodium hydroxide by naming it 'sodum isethionate'.

## *Did you know? About detergents*

*Index 09.10*

The information can be given to students while they are waiting for trace to occur, or as part of the lesson on soap-making if this is done as a demonstration.

### Answers

1. Students may give various answers, dependent on the carbon chain length. An example of an equation would be:

    coconut + sodium hydroxide → glycerol + sodium myristate
    sodium laurate
    sodium palmitate

The molecular structures are also given for teachers' reference:

$$H_2C-O-\overset{\overset{\displaystyle O}{\|}}{C}-(CH_2)_{12}CH_3$$
$$HC-O-\overset{\overset{\displaystyle O}{\|}}{C}-(CH_2)_{10}CH_3 \; + \; 3NaOH \longrightarrow$$
$$H_2C-O-\overset{\overset{\displaystyle O}{\|}}{C}-(CH_2)_{14}CH_3$$

$$H_2C-OH$$
$$HC-OH \; +$$
$$H_2C-OH$$

$$^+Na^-O-\overset{\overset{\displaystyle O}{\|}}{C}-(CH_2)_{12}CH_3$$
$$^+Na^-O-\overset{\overset{\displaystyle O}{\|}}{C}-(CH_2)_{10}CH_3$$
$$^+Na^-O-\overset{\overset{\displaystyle O}{\|}}{C}-(CH_2)_{14}CH_3$$

2. The names of possible soaps which can be made from coconut oil are: sodium myristate, sodium laurate, sodium palmitate, sodium caprylate, sodium caprate and sodium stearate.

   It is possible that soap bars may include sodium myristate, sodium stearate and sodium palmitate. The more expensive bars are likely to contain sodium palmitate and possibly myristate. Cheaper soaps may contain sodium stearate, but this will be from animal fats. Soaps based on animal fats tend to cost less than those made from plant oils. You could discuss the benefits - are more expensive soaps kinder (less drying/irritating) to the skin? What is the evidence for this?

3. Soap must be left to cure because the reaction is incomplete immediately after the soap is poured into moulds.

4. The energy available at room temperature is sufficient to generate a slow rate of reaction.

5. Increasing the temperature would increase the rate of reaction. This would produce the soap more quickly, so the manufacturer would produce more soap per day. However, the oil molecules may be broken down at higher temperatures, destroying the soap compounds. There has to be a balance.

6. When soap detergents are used in hard water areas a solid scum forms due to calcium and magnesium ions in the water reacting with the soap. However, due to the different composition of soapless detergent, the solid substance that forms dissolves in water leaving no scum. Water softeners are added to some products in order to reduce the amount of scum produced.

## *The history of soap*

*Index 09.11*

Learning objectives
• To understand how the way we live in modern society has developed.

Time required
20 minutes

Answers

1. The chemicals used to make soap are animal fats, tallow, whale fat (blubber), olive oil, wood ash.

2. An example might be:
   Soap was taxed because of the **competition** for tallow. Poor people needed to light their houses. Tallow was used for candles. Tallow was in high **demand** for this and soap-making. If tallow was used for soap, there was less for candles. This made candles more expensive. To protect tallow for candles, soap was taxed and made more expensive instead.

3. The tax was lifted; plumbing improved; water supplies improved; people started bathing regularly; lighting was based on gas not candles so tallow could be used for soap.

4. Using soap made them feel clean and smell much better.

5. The soap was probably very rough in quality - maybe had bits of wood in it; not very pure; not very hygienic.

6. Soap was linked to being clean; people began to realise the link between being dirty and dirt carrying bacteria/viruses which made people ill. The manufacturers tried to play on this, claiming that their soap would help keep the users healthy. The name 'Lifebuoy' is like claiming the soap 'rescues' the user from a dirty way of life.

7. All these soaps are made today, but they are slightly different to how they used to be! For example, Wright's Coal Tar soap no longer contains phenol (carbolic acid), but instead 'Coal Tar Fragrance' is used. Note that the soap was called 'Coal Tar' because carbolic acid was extracted from coal tar, the residue left behind from distilling coal.

## *Using soaps and soapless detergents*

*Index 09.12*

Learning objectives
• To rationalise different scientific views.

Time required
30 minutes

In my opinion...

The activity is self-explanatory but some useful tips might be:
- Arrange students in groups of about 4 - not too many or the discussions will be uncontrollable.
- Use laminated A3/A2 sheets of card for feedback - give one sheet to each group with a dry-wipe marker pen. They can show the ranking and write down the reasons. If they change their mind they can wipe off and start again. The cards also make a good way of displaying the ideas - ask one group member to show the card to everyone else.
- Remind students to use the chemistry they have learned in the other activities.
- The student statements could be given out on strips of paper so groups can arrange these in a rank order easily.

## Forever blowing bubbles

*Index 09.13*

Learning objectives
- To design an experiment to investigate the action of soap on bubble bath.

Time required
60 minutes

The actual answer to the question is that both bubble bath and soap contain surfactants. They are different types of surfactant. Bubble bath contains cationic detergents, which make good foams. Soaps contain anionic molecules which have good washing characteristics.

When soap and bubble bath meet, the two oppositely charged surfactant molecule heads are attracted to each other, creating a complex. The complex is neither good for foam nor cleansing, as the properties of both are lost.

When taking a bath, the soap is usually used in excess, so once all the bubble bath has reacted with the soap then cleaning can take place.

First students would need to test the effect, investigating the quantities of bubble bath and soap involved. Then the variables of bubble bath and soap could be investigated.

## Further information, references and source material

Websites for information and ingredients for making soap and detergents include:

*www.nealsyardremedies.com* (accessed November 2003)

*www.thesage.com* (accessed November 2003)

*www.coconutoil-online.com* (accessed November 2003)

*www.snowdriftfarm.com* (accessed November 2003)

M. Coss, *The handmade soap book*, London: New Holland, 1998.

M. Davies, *Rough Science*, Milton Keynes: The Open University, 2000.

Neal's Yard Remedies, *Natural Health and Body Care*, London: Aurum Press, 2000.

Neal's Yard Remedies, *Make your own cosmetics*, London: Aurum Press, 1997.

M. O'Hare, ed. *The Last Word*, Oxford: Oxford University Press, 1998.

B. Selinger, *Chemistry in the Marketplace*, 5th edition, Sydney: Harcourt Brace, 1998.

C. H. Snyder, *The Extraordinary Chemistry of Ordinary Things*, 3rd edition New York: John Wiley, 1998.

R. Winter, *A consumer's dictionary of cosmetic ingredients*, New York: Three Rivers Press, 1999.

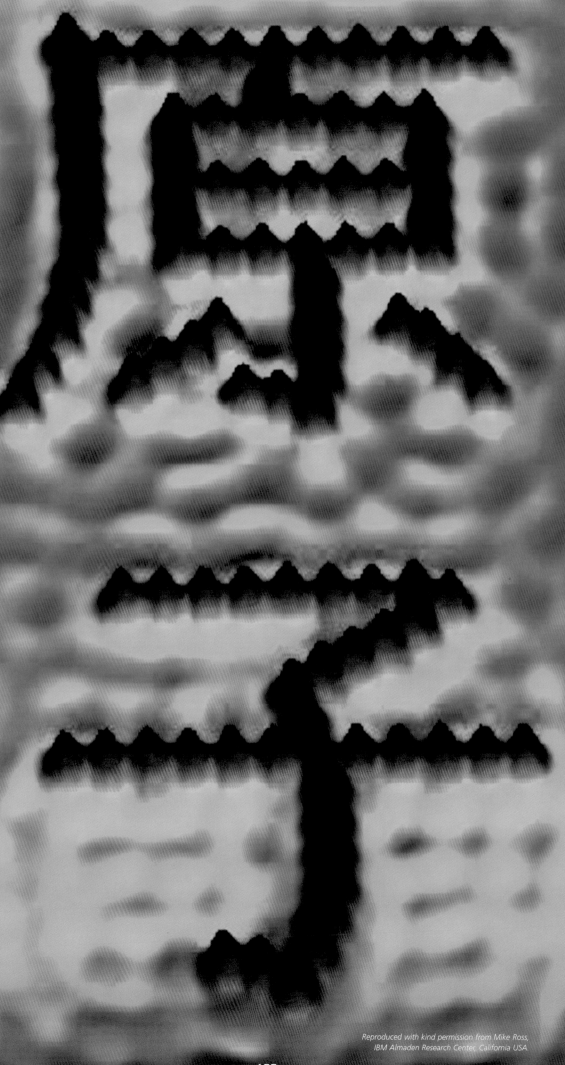

Nanochemistry

Reproduced with kind permission from Mike Ross,
IBM Almaden Research Center, California USA.

**Summary**

These resources explore atoms and nanotubes and how they are being used and developed for use in everyday life.

| Resource name | Index | Type | Age range | Topic | Media |
|---|---|---|---|---|---|
| Introduction to atoms | 10.01 | Paper based exercise. Interactive exercise: The smallest thing... | 11–16 | Size | |
| Did you know? Getting down to nanometres | 10.02 | Information | 11–16 | Units | |
| Demonstration: Atomic paper clips | 10.03 | Demonstration and questions | 11–14 | The nature of atoms | |
| Tasty particles | 10.04 | Class experiment and questions | 11–14 | Size of atoms | |
| Did you know? About the Scanning Tunnelling Microscope (STM) | 10.05 | Information | 14–16 | Instrumentation | |
| Seeing atoms | 10.06 | Literacy / DART and questions Interactive exercise | 14–16 | Atoms | |
| Did you know? How STM pictures are made | 10.07 | Information | 14–16 | STM images | |
| Carbon and its various forms | 10.08 | Literacy / DART | 14–16 | Allotropes | |
| Modelling nanotubes | 10.09 | Class paper activity and questions | 14–16 | Nanotubes | |
| Combining nanotubes | 10.10 | Class paper activity and questions | 14–16 | Nanotubes | |
| The challenge to silicon | 10.11 | Literacy / DART | 14–16 | Molecular computers | |
| Key words | 10.12 | Glossary | 11–16 | Glossary | |
| The nanoscience conference | 10.13 | Poster activity – communication | 14–16 | How scientists present their work | |
| There's plenty of room at the bottom | 10.14 | Literacy and questions | 14–16 | The nature of science | |
| Using Euler's Law and polyhedra | 10.15 | Interactive activity | 11–16 | Euler's Law | |

Key:   Interactive student activity    Photocopiable and printable worksheet    Projectable picture resource

Issue
Should we be frightened of nanotechnology? Or is the hype just that?

Chemical topics
- What are the relative sizes of atoms?
- What is an element?
- Can we see atoms?
- How big are atoms?
- Nanostructures and improvements in technology.

Scientific enquiry issues
- Limitations of models
- Mathematical relationships
- Making scientific predictions.

Notes on using the resource

The unit begins by looking at atoms – students are invited to develop a sense of the small size of atoms and then to look at STM pictures of atoms. After this, activities present nanotubes – tiny tubes of carbon atoms, showing what these look like through modelling them. In **The challenge to silicon**, students can find out how nanotubes are being used to make electric circuits and potentially the next generation of computers. Exploring a variety of potential developments is the theme of **The nanoscience conference**, at which students are invited to present a poster at a poster session run like a real scientific conference. Finally, in **There's plenty of room at the bottom**, part of a Richard Feynman lecture is used to help students appreciate the role of vision in developing science.

## *Introduction to atoms*

*Index 10.01*

Learning objectives
- Atoms exist but are very small – smaller than we can see or imagine
- Atoms and molecules can diffuse.

Time required
About 30 minutes

This activity is designed to get students thinking about things they consider as 'small', then to put atoms in perspective against these.

Alternatively teachers may choose to use an analogy to something within the students grasp to convey the concepts of large numbers and of how small atoms are. For example, Show a 1 kg bag of risotto rice – 25 g of rice (raw) is approximately 1000 grains, therefore, $10^6$ grains is approximately

25 kg of rice. How many grains of rice in the 1 kg bag? Answer: about 40,000 – therefore, about 25 x 1 kg bags of rice would contain about 1 million grains.

The items to go in the suggested table could be generated by a class brainstorm, or students could think of these individually. Possible items include:

Items that can be seen:
- 0.5–3 mm    pin tip, sand grains, specks of dust, tiny flies or other insects, holes/weaving in fine material;
- 3–5 mm    insect larvae, creases in skin, shapes of fingerprints, crystals of salt and sugar, drops of liquid, eyelashes.

Items that cannot be seen:
- 0.1 mm    details of salt crystal shapes, mites, details on a silicon chip;
- 10–100 μm    cells, bacteria (variation here depending on species);
- 0.3–1 μm    bacteria, viruses, DNA;
- 1 nm    molecules, atoms;
- 1 pm    atomic nuclei;
- 0.1 fm    the matter in a proton.

Once students have generated a list, some research would give the size information needed. A good source is *Powers of Ten* by Philip Morrison and Phyllis Morrison (see resource list). This was used to provide the information here. **Getting down to nanometres** discusses the measurements. The interactive activity **Using Euler's Law and polyhedra** could help in giving examples.

One nanometre is 0.000000001 m. It can be written as 1 nm or $1 \times 10^{-9}$ m. Here is the scale of length showing where nanometres fit in:-

| Small | attometre | am | 0.000000000000000001 m | $1 \times 10^{-18}$ m |
|---|---|---|---|---|
| | femtometre | fm | 0.000000000000001 m | $1 \times 10^{-15}$ m |
| | picometre | pm | 0.000000000001 m | $1 \times 10^{-12}$ m |
| | nanometre | nm | 0.000000001 m | $1 \times 10^{-9}$ m |
| | micrometre | μm | 0.000001 m | $1 \times 10^{-6}$ m |
| | millimetre | mm | 0.001 m | $1 \times 10^{-3}$ m |
| | centimetre | cm | 0.01 m | $1 \times 10^{-2}$ m |
| | metre | m | 1 m | $1 \times 10^{0}$ m |
| | decametre | dm | 10 m | $1 \times 10^{1}$ m |
| | hectometre | hm | 100 m | $1 \times 10^{2}$ m |
| | kilometre | km | 1000 m | $1 \times 10^{3}$ m |
| | megametre | Mm | 1000000 m | $1 \times 10^{6}$ m |
| | gigametre | Gm | 1000000000 m | $1 \times 10^{9}$ m |
| Large | terametre | Tm | 1000000000000 m | $1 \times 10^{12}$ m |

The metre is the standard (SI, or Système International d'Unitiés) unit of length. Every other unit is stated as a number bigger or smaller than this. The short word put before metre is called a prefix. Many of these are from Greek. The same prefixes are used to change the unit of mass, the kilogram, into smaller and larger units. Atoms and molecules are nano- and picometre-sized. Science involving nano-and pico-sized particles is called nanoscience.

Answers

1. See their table for items. We cannot see below about 0.5 mm length.

2. Again, this depends on the table, but the most accurate response is the matter in subatomic particles at about 0.1 fm ($1 \times 10^{-16}$ m).

3. Our eyes are not sensitive enough.

4. The small things are made from atoms, which in turn are made of matter. No-one really knows what this looks like.

5. Atoms are very small!

## *Did you know? Getting down to nanometres*

*Index 10.02*

## *Demonstration: Atomic paper clips*

*Index 10.03*

Learning objectives
• Atoms are the smallest parts of elements
• An atom retains the property of the element.

The demonstration shows that the smallest part of any substance which can be identified as the substance is an atom. One paper clip represents an atom of paper clips. The demonstration also implies that atoms must be very small.

## Time required
About 10 minutes

## Apparatus and equipment
- A pile of paper clips – a box of 100 small ones tipped out will suffice
- A sheet of fool's gold (*eg* as used in an electroscope) or aluminium foil.

## What you do
1. Place the pile of paper clips and the gold/aluminium side by side.
2. Start with the paper clips.
3. Divide the paper clip pile into roughly half, making two smaller piles.
4. Repeat the division until there are two paper clips and then finally, to one paper clip alone.
5. The single paper clip represents an atom.
6. Now split the paper clip into two pieces. Invite names for these, such as 'hook' and 'centre'.
7. Now turn to the gold/aluminium sheet.
8. Tear the sheet into two halves.
9. Repeat the tearing until the pieces cannot be torn any smaller.
10. As the piles and gold/aluminium are torn apart, ask students to answer the questions on the sheet. Answers are given.

## Answers

1. One paper clip.

2. Yes.

3. Names for the paperclip might be hook, centre. Parts of an atom are proton neutron and electron.

4. No, the paper clip is not useable. A broken up atom would not be useable as an atom.

5. An atom of gold/aluminium.

6. We can't use our hands to tear pieces that small.

7. Yes.

8. We need to use a very powerful microscope.

9. Yes.

10. No.

# RS•C

## Nanochemistry

## *Tasty particles*

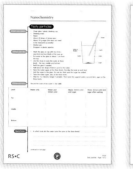

Index 10.04

Learning objectives
• Diffusion is a kinetic effect
• Particles are very small.

Time required
About 30 minutes for a class experiment; 15–20 minutes as a demonstration. This experiment shows that particles move – but their small size prevents us from seeing this movement. Particle movement is called diffusion – here, particles of lemon juice (citric acid) and sugar $C_6H_{12}O_6$ are added to water and diffuse at different rates. This can be discussed in relation to relative particle size. The experiment could be done either as a demonstration with student assistance, or as a class demonstration. As the experiment involves tasting, care should be taken that a class experiment is done in a clean (*ie* non laboratory) environment.

Note – teachers may like to use this exercise as an example of comparing like with like; sugar involves dissolving and diffusion whereas lemon juice only involves diffusion.

Apparatus and equipment (per demonstration or group of students)
• 1 glass or plastic drinking cup
• 1 drinking straw
• Tap water
• About 20 drops of lemon juice – depends on the lemon, but sufficient to taste
• About 10 g sugar (no need to weigh accurately: 1–2 teaspoons is sufficient)
• Marker pen
• 1 dropper or plastic pipette
• Eye protection.

What you do
1. Mark the glass or cup with two lines – one-third and two-thirds of the way up.
2. Pour water into the glass to about 1 cm from the top.
3. Use the straw to taste the water at three levels– the top, middle and bottom.
4. Refill the glass if needed.
5. Add about ten drops of lemon juice to the water.
6. Taste the water again at the three levels. Describe the taste at each level.
7. Add the sugar to the glass. Do not stir. Allow the sugar to settle.
8. Taste the water again at the three levels.
9. Wait for 1–2 minutes. Then taste the sugared water a second time, again at the three levels. The sugar taste takes a few minutes to spread – if possible, wait longer and then taste again. Encourage observation of the sugar – this will dissolve slowly.

161

### Results

| Level | Water only | Water and lemon juice | Water, lemon juice and sugar | Water, lemon juice and sugar after waiting |
|---|---|---|---|---|
| Top | No taste | Lemon | Lemon | Lemon |
| Middle | No taste | Lemon | Lemon | Lemon and sugar |
| Bottom | No taste | Lemon | Lemon and sugar | Lemon and sugar |

### Answers

1. Water only, water and lemon juice.

2. Water, lemon juice and sugar before and after waiting.

3. The sugar molecules took more time to move through the water as they had to dissolve.

4. The particles are not the same. Otherwise, when sugar is added, the water would taste the same immediately at all three levels, like when the lemon juice is put in.

An alternative experiment that is a little more spectacular involves students reacting lead nitrate and potassium iodide by adding a crystal of each to a petri dish of deionised water. Full details of this experiment can be found in K. Hutchings, *Classic Chemistry Experiments*, London: Royal Society of Chemistry, 2000, pages 68-69.

## *Did you know?  About the Scanning Tunnelling Microscope (STM)*

*Index 10.05*

## *Seeing atoms*

  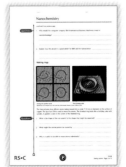

*Index 10.06*

This activity reinforces ideas about atoms.

### Learning objectives
- Atoms can be 'seen' using a scanning tunnelling microscope (STM)
- Atoms are nanometre-sized particles.

### Time required
About 30 minutes

In this activity, students will look at pictures of atoms taken using an STM, answering questions about each picture. The aim is to encourage students to think about the scale of the pictures – atoms are nanometre sized. Use **Did you know? How STM pictures are made** to provide perspective.

### Answers

*The zit*
*Reproduced with kind permission from Mike Ross, IBM Almaden Research Center, California USA.*

1. Nickel: symbol Ni, atomic number 28, mass number 59
   Xenon: symbol Xe, atomic number 54, mass number 131

2. The xenon atom has many more sub-atomic particles or electron shells than nickel.

3. No, the colours have been added by the computer.

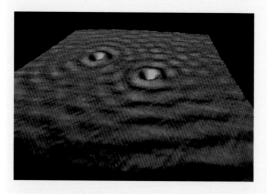

*The dents*
*Reproduced with kind permission from Mike Ross, IBM Almaden Research Center, California USA.*

1. Like waves.

2. Water or perhaps light.

3. Electrons can behave like waves.

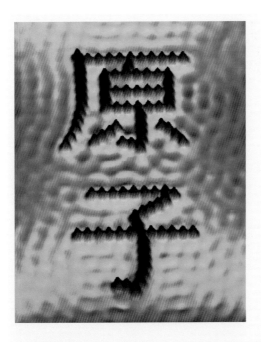

*An 'original child'*
Reproduced with kind permission from Mike Ross, IBM Almaden Research Center, California USA.

1. A count reveals that 100 iron atoms have been used in the picture.

2. The distance will be 248 x 11 x $10^{-3}$ = 2.728 nm.

3. 'Original child' reflects the sense of atoms as the basic building blocks for other substances.

*In the beginning...*
Reproduced with kind permission from Mike Ross, IBM Almaden Research Center, California USA.

1. IBM is a computer company which will always be seeking to develop new technologies and nanotechnology may provide faster computers.

2. It showed that IBM could lead in this area by demonstrating what could be done by moving atoms.

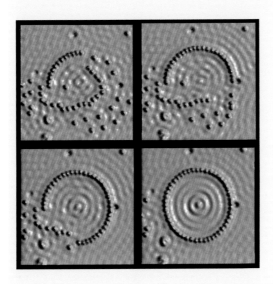

*Ironing the prefect circle*
Reproduced with kind permission from Mike Ross, IBM Almaden Research Center, California USA.

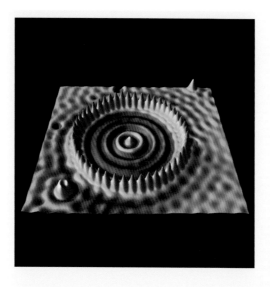

*The birthday cake*
*Reproduced with kind permission from Mike Ross, IBM Almaden Research Center, California USA.*

1. They are spikes. Most students think of atoms as spheres, so the answer to this will probably be no.

2. Electrons.

3. Scientists can make the kinds of molecules they really want without having to mix chemicals.

  ## *Did you know? How STM pictures are made*

 *Index 10.07*

 ## *Carbon and its various forms*

 *Index 10.08*

Learning objectives
• To find out about new giant structures of carbon
• To model nanotubes and appreciate the role of modelling in chemistry.

# Nanochemistry

## Time required
40 minutes; 60+ minutes if extension work is included

This activity introduces a new and developing aspect of nanoscience – nanotubes. Students begin by finding out what nanotubes are. Then they make a model of a nanotube – there is no hidden reason for doing this, other than to model something which is very small. Making models often helps scientist get a view for what might be happening at an atomic level - the base pair arrangements in DNA and the proposed $C_{60}$ (buckyball) structure were famously modelled first in paper!

## What do nanotubes look like?

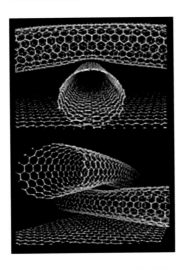

*Piled high*
Reproduced with kind permission from Mike Ross, IBM Almaden Research Center, California USA.

## Answers

1. Students may say that the nanotubes look like nets; hexagon shapes, flexible.

2. Carbon.

3. Graphite. The nanotubes pictured here are single hexagon layers rolled up. The graphite structure is a stack of hexagonal layers.

4. The electron arrangement in a carbon atom allows formation of a maximum of four covalent bonds. Provided this requirement is satisfied different arrangements are possible.

 ## *Modelling nanotubes*

  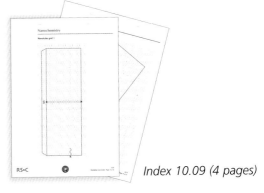

*Index 10.09 (4 pages)*

# Nanochemistry

Learning objectives
• Use of models to visualise structures.

Time required
40 minutes

The point of this task is to make models of nanotubes. This will help students appreciate the small size of atoms and the place of modelling in chemistry.

Notes on nanotubes and on making the models

1. There are basically two types of nanotube – called metallic and semiconducting. Grid 1 produces a (15,0) metallic nanotube, while grid 2 produces an (11,7) semi-conducting nanotube. The numbers indicate the carbon atoms joined across the diameter of the tube. The (15,0) nanotube has 15 carbon atoms joined in a straight line at right angles to the tube axis. The (11,7) nanotube has the line of carbon atoms tilted at an angle of approximately 10°. If times allows, teachers may suggest that students make both types so they can appreciate the difference. Otherwise, ensure the two grids are distributed evenly around the group.

2. Nanotubes are metallic when the difference between the numbers is a multiple of 3. This can also happen when both numbers are the same and are a multiple of 3, such as in the (6,6) nanotube.

3. The differences in numbers of carbon atoms leads to nanotubes varying in diameter. In making longer tubes, it is therefore best if tubes of the same type are joined together to ensure a good 'fit', hence the need to ensure plenty of each. Joining ten to make a 1 m long model works well.

4. The scale of the models is 1 cm = 0.5 nm, or 2 cm = 1 nm. This corresponds to the diameter of the grid 1 and 2 models. The 10 cm length = 5 nm. Exceptionally, nanotubes can be up to 100 mm (100 000 nm) long. This means up to 20 000 would be needed to model a real tube length – this is equivalent to 2 km.

5. Both semi-conducting and metallic nanotubes can be chiral. This happens when the line of carbon atoms is at an angle. Connecting about 10 (11,7) tubes together using the marked lines as a guide shows that the line winds around the tube, creating a repeating twist running in one direction and hence chirality, as the twist could be reversed in a 'partner' tube.

6. Making the nanotubes can be difficult. They are made using laser vapourisation of carbon at 1200 °C. Single-walled and multiple walled tubes are possible. Techniques for making nanotubes are developing – at the moment, the type of tube cannot be confidently predicted.

7. The length of the tubes confirms the stability of the structure and hence the relative strength of carbon-carbon bonds. If the nanotubes were available on a macroscale, they would be stronger than a steel rope.

8. Remember that this field is a developing one. The facts provided are correct at the time of writing, but please check information independently for the latest developments.

# Combining nanotubes

## Learning objectives
• Use of models.

## Time required
30 minutes

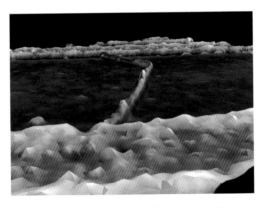

Figure 1a A 'kinked' nanotube joining electrodes (yellow) on a silicon dioxide base (green)

Reproduced with kind permission from Cees Dekker, TU Delft, The Netherlands.

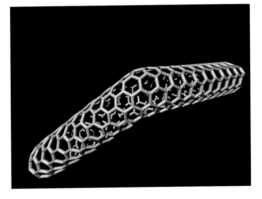

Figure 1b A model of the nanotube joining the two electrodes

The model shows a metallic (top part) and a semiconducting nanotube joined together.

To make the 'kink', the tubes need to be joined at an angle. This leaves a space at the join which needs to be filled. The picture shows that this is best filled with a pentagon shape. This would be tricky to model on this scale, but can be discussed.

## Answers

1. The mathematical link between the numbers controls if a nanotube is metallic or semiconducting. This is the main difference.

2. The nanotubes could be used to make electronic circuits. They could also be used as tiny wires or strings to strengthen other materials in a composite.

3. The diameters of the finished tubes are different and depend on the length of the line drawn between the two rows of hexagons.

4. The connection is that all the numbers in the metallic nanotubes are multiples of three. Mathematically, $i-j=3x$.

**Getting longer...**

5. Ten model nanotubes joined together makes 1 m. This on the scale of 1 nm = 2 cm is equivalent to 50 nm. This is 50,000 μm.

6. $2 \times 10^8$ cm.

7. $2 \times 7^{10}$ nanotubes.

## *The challenge to silicon*

*Index 10.11*

**Learning objectives**
- To understand how nanotubes can be used to develop a new generation of computers
- To understand that technologies compete for new developments.

**Time required**
30 minutes

**Answers**

1. They will be much smaller and made using nanowires that can be compared to the size of molecules – they are nanometre-sized.

2. Nanotubes have smaller diameters but longer lengths than nanowires, which are shorter and fatter. At this scale, the size difference will not matter to us, because we cannot see these measurements. Computers made in both ways would appear tiny to us.

3. Nanotubes need only one element which is readily available. Nanowires can be made more easily than nanotubes.

4. Using nanotubes – making a good, predictable supply: both methods - making enough connections to build up a circuit.

5. Breaking a size barrier is intellectually satisfying and a practical challenge; using smaller items saves on the world's resources; smaller items take up less room; usually smaller items work more efficiently, so saving time.

## *Key words*

*Index 10.12*

 ## *The nanoscience conference*

*Index 10.13*

### Learning objectives
- To research an aspect of nanoscience
- To learn how scientists present research
- To consider how scientific developments may impact on life.

### Time required
Variable – minimum 60 minutes for research and 60 minutes for the conference. Extra time will be needed to prepare the conference itself.

This activity is designed to imitate a real session at a scientific conference. Students are encouraged to research a topic and then to present a poster under the given constraints. This is an exciting opportunity to present new developments in a potentially key area of chemistry, bordering too on physics and biology. Points to bear in mind in running this activity include:

1. Prepare the notice – the basic invitation can be adapted to suit the situation by filling in the date, time and place Add other information to the invitation – for example, names of invited guests (see below) and pictures of nanotubes to create an exciting and positive image for the event.

2. Allowing ample time for the research to be completed. The information on the websites may need deciphering if it is written for a scientific audience. This could be an on-going project for several weeks during a topic on particle ideas, or structures, with the conference as the culmination.

3. Ensure all posters are displayed before the time the conference is due to start.

4. Encourage students to take turns standing by their posters – maybe change after a few minutes so everyone gets the chance to circulate and see others' work.

5. Invite guests – such as older students (17-18 year olds), senior (non-scientist) colleagues, colleagues from a university chemistry department – to the conference. This always adds an extra 'buzz', making the students feel their efforts are recognised elsewhere.

6. Be flexible and don't worry about not knowing 'the answers'- this is a developing field in which hard and fast facts are relatively few and are evolving. Students need to realise that not all chemistry comes directly from a textbook and that new developments mean exactly that!

7. Experience of this type of event suggests that a competitive 'edge' is also helpful in creating a good atmosphere – consider ways of rewarding particularly good efforts in design, presentation and content of the posters if this is appropriate in your setting.

8. Keep the posters for display at open evenings and parents' events.

Projects and related websites

Quantum dot nanocrystals: *www.qdots.com*  (accessed November 2003)
Catalysts: *www.bza.org/zeolites.html* (accessed November 2003)
Nanoshells: *www.ece.rice.edu/~halas/*  (accessed November 2003)
Making nanotubes: *www.cnanotech.com* (accessed November 2003)
Nanotube drugs to kill bacteria: *www.scripps.edu/chem/ghadiri/html/research.html* (accessed November 2003)
Nanosized electric circuits: *www.mb.tn.tudelft.nl/nanotubes.html* (accessed November 2003)
Transparent sunblock: *www.nanophase.com/* (accessed November 2003)

By the time you do this, these projects may have been replaced by others – nanotechnology will not disappear! The point is to include a range of different aspects of the field. There is no reason why two groups of students may not tackle the same subject.

Discussion: Back to the future

• This is a good opportunity to enable students to see the potential in scientific discoveries and developments. If they have prepared their conference posters and discussed the projects, they should have suggestions to make in this discussion.

• Allow about 20 minutes and remember it is best to end while they are 'wanting more', rather than allowing discussion to die. Consider a follow up homework to summarise the key points.

• If guests attend the conference, try to have the discussion with one or more present and ascertain their views too.

• There is potential to develop this further into a News article, perhaps for a school magazine, even a local paper report.

# 1st Nanoscience Conference

at (Name of school or college)

The conference theme is:

# New uses for Nanoscience

You are invited to take part in a

# Poster Session

to be held on (date)

at (time)

in (place)

Posters must be A2 size (4 x A4 sheets) and include:
●
the project title
●
the names of the scientists working on the project
●
an explanation of how the project uses nanoscience
●
why the project is useful or important
●
how the project might develop in the future
●

## *There's plenty of room at the bottom*

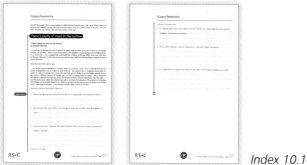

Learning objectives
- To consider how vision influences the development of science.

Time required
30-40 minutes

This activity uses a short extract from a talk by Richard Feynman, who was regarded as one of the key figures in physics in the 20th century, winning a Nobel Prize in 1965. He presented a vision in this talk, now thought of as a 'classic', showing a way forward for science. We cannot all be a Feynman, but you never know – there may be another one in your class waiting to be prompted. Students may enjoy the chance to find out more about Feynman's work.

More information about Richard Feynman can be found at:
*www.feynmanonline.com* (accessed November 2003)
*www.zyvex.com/nanotech/feynmanWeb.html* (accessed November 2003)

Teachers may like to use the following article that appeared in *Education in chemistry*, 40(4), 2003.

## First steps into a small world

Among some of the most notable scientific discoveries of the past century that have made a major impact on our society include the silicon chip, DNA, biotechnology, and the human genome. If worldwide government spending and company investment – and recent media attention – are a gauge, then the next few decades look set to witness another. Billions of dollars are being ploughed into R&D in 'nanoscience' or 'nanotechnology' mainly in the US, Japan, China, Western Europe, Australia and Korea. What is essentially the exploitation of fundamental physics and chemistry, coupled with an insight into the way Nature works and builds materials, nanotechnology holds great expectations. So what is nanoscience, and where is it likely to impinge on our society?

Enter Richard Feynman...

A nanometre is one billionth of a metre, ie $10^{-9}$ m. To put this into some sort of perspective, a human hair is ca 20000 nm in diameter, a microbe is *ca* 200 nm in diameter, and double-stranded DNA is *ca* 2.4 nm in diameter. Nanoscience refers to structures that are 1–100 nm in size, roughly of the order of a few hundred atoms/molecules. Structures of this dimension have been around for a long time, what's different today is the ability of scientists to see and control atoms at this dimension.

The story began back in 1959, when Richard Feynman, the US Nobel prizewinning physicist gave a talk to the American Physical Society, entitled 'There's plenty of room at the bottom'. In what has become a famous speech, during which he described, in theory, a way of putting the entire 24 volumes of Encyclopaedia Britannica on the head of a pin, he asked, 'what would the properties of materials be if we could really arrange the atoms the way we want them?'. 'I can hardly doubt', he said, 'that when we have some control of the arrangement of things on a small scale, we will get an enormously greater range of possible properties that substances can have, and of different things we can do'. By way of example he cited glass and plastics, which on the macroscale are amorphous solids, but on the nanoscale have structures that are more ordered.

It took another 20 years, however, for scientists to come up with techniques that would allow them to see and move individual atoms on molecular substrates. In the early 1980s, Gerd Binning and Heinrich Rohree, working at IBM's Research Centre in Switzerland, published their work on a 'scanning tunnelling microscope (STM). The STM essentially uses an atomic size tip (usually made of tungsten) to move across a surface; electrons tunnel their way from a group of atoms on the surface, producing a very small current, which can be recorded and displayed on a computer. Using the STM, the physicists were able to construct enlarged images of surfaces, allowing them to 'see' atoms and molecules for the first time. In 1986 they received the Nobel prize for physics for this work. By the early 1990s other scientists had found ways to isolate and move atoms by using related techniques and for doing structural analysis at this level, with much of this work being recognised with Nobel prizes by the mid–late 1990s.

### New things in small packages

It turns out, as Feynman predicted, that nanoparticles exhibit different behaviour when their sizes fall below a critical length associated with any given property. Some of these new properties arise because 'quantum phenomena', ie the properties of atoms and electrons and how they affect matter; become important on the nanoscale. Electrons, for example, can now move through an insulating layer of atoms, which cannot be explained by classical physics. In general scientists, have found that nanoparticles are more sensitive to light, heat and other physical forces such as magnetism and electrostatics than the same particles in the bulk.

Another characteristic of nanoparticles is their high surface area to volume ratio, compared with conventional materials. As a material gets smaller, the proportion of their constituent atoms at or near the surface increases. Nanostructures with well-defined pore sizes and high surface areas are being investigated for use in energy storage and separation processes such as $H_2O$, $H_2S$ and $CO_2$ removal. Others are working to produce highly selective sensors, new coatings, drug delivery systems, and stronger and lighter construction materials whose bonding and strength depend on the surface area and the shape of the nanoscale structures. Nanostructures are also expected to have a significant impact on the development of new product-specific catalysts. Chemists in Japan, for example, have discovered that gold particles smaller than 3–5 nm show catalytic behaviour; deactivating noxious CO at temperatures as low as $-70^0C$ (and converting it to $CO_2$). Normally this reaction takes place at much higher temperatures. Interestingly, they also discovered that these nanoscale gold particles (crystals) have a very different arrangement to gold particles in the bulk.

### Beyond the atom

Despite the recent scare stories in the press – *eg* of 'the "grey goo" that could result from the uncontrollable spread of self-replicating nanoscale machines' – nanotechnology currently lies mainly in the realms of the R&D laboratories, with scientists working hard to find out just what novel properties can be found for myriad materials on this scale. Most of the applications are still many years away. There are a few exceptions, however. Much work has already been done over the past 15 years in the area of nanocomposite materials where nanostructures are introduced into other materials. Nanocomposite polymers that are stronger and stiffer than ordinary plastics, are already being used to replace metals in the car industry, while other nanocomposites are being developed with medical and dental applications in mind. Nanoparticle fillers in metal, ceramic and polymer layers are expected to yield a variety of nanocomposites with unique properties.

Other applications already on the market include:

ICI's latest sun creams – these use titanium dioxide ($TiO_2$) nanoparticles of *ca* 50 nm diameter in contrast to the older products which use $TiO_2$ pigment particles which are *ca* 200 nm in diameter. The former absorb light and scatter much less than the larger particles. Consequently, the new product, unlike the older one, leaves no white smears on the body;

BASF use $TiO_2$ nanoparticles (*ca* 500 nm diameter) in textile fibres (polymer composites). The materials are used to absorb UV radiation in sun protective clothing.

Finally, it is worth mentioning one other area of current intense research activity. Many scientists believe that carbon nanotubes, will prove to be nanotechnology's answer to replacing silicon in a host of smaller and faster electronic devices. Discovered over 10 years ago by the Japanese chemist Sumio Iijima while working on fullerenes ($C_{60}$, $C_{70}$, …), carbon nanotubes are rolled up cylinders of graphite with diameters *ca* 1 nm and reaching lengths of micrometres. They are thought to be stronger than steel, and depending how the graphite molecules line up they can be semiconducting or conducting. A current drawback to the exploitation of these structures is the fact that synthesis leads to a mixture of carbon nanotubes with varying lengths and orientations. Research in this field is focusing on ways to select structures with a specific size and orientation. DuPont chemists in the US, for example, have found that they can use DNA to bind to specific carbon nanotubes, offering them the potential to separate them. The race for the molecular computer, however, is still wide open.

Thus nanotechnology holds much promise for the future, though the general consensus among scientists is that progress will be evolutionary rather than revolutionary.
Kathryn Roberts

Answers

1. Feynman was talking about the world of atoms and subatomic particles.

2. The talk was given over 40 years ago and presents a vision for a future – no-one else was seeing the possibilities in this way; also what Feynman was talking about has happened to a large extent.

3. The chemist's way of mixing chemicals, shaking and fiddling around; the atomic scientist's way of picking up atoms and placing them where the chemist wants.

4. Yes, it is possible. This might help because molecules can be made to specific requirements.

5. Yes, to at least some extent it is real, and the possibilities are being worked out. Scientists can now manipulate atoms as they want to.

6. Yes, it is important for scientists to have a vision, because this helps take the subject on further and brings in new areas.

 ***Using Euler's Law and polyhedra*** *Index 10.15*

This is an interactive exercise based on Euler's Law and introduces students to mathematical and scientific relationships.